LITTLE FACTS OF LIFE

LITTLE FACTS OF LIFE

350 Mini Readings in Biology

by

Eddie Lunsford

iUniverse, Inc.
Bloomington

Little Facts of Life
350 Mini Readings in Biology

iUniverse books may be ordered through booksellers or by contacting:

iUniverse
1663 Liberty Drive
Bloomington, IN 47403
www.iuniverse.com
1-800-Authors (1-800-288-4677)

ISBN: 978-1-4759-7770-7 (sc)
ISBN: 978-1-4759-7771-4 (ebk)

Printed in the United States of America

iUniverse rev. date: 03/09/2013

CONTENTS

CHAPTER ONE

KINGDOMS MONERA, PROTISTA & FUNGI

Bird's Nest Fungi

They don't grow on bird's nests. Rather, these unusual fungi have reproductive bodies (called peridia) that, with a little imagination, resemble a small bird's nest complete with tiny eggs. Six genera are known, all belonging to the order Nidulariales. The "bird nest" is actually a small cup-like structure especially designed to catch raindrops and funnel them downward. Within the nest are small, egg-like structures known as peridioles. They contain spores with the ability to germinate and develop to form new organisms. The force of the raindrop causes the peridioles to be forcefully ejected out of the nest and into the environment, thereby spreading the species. In some genera these peridioles are sticky. Others look like tiny mushrooms and have a small thread, the funicular cord, which may wrap around any object in the path of the periodiole as it is ejected.

Sour Bacteria

If you enjoy the taste of vinegar as a flavoring in various foods, you are indebted to an organism known as *Acetobacter aceti.* This tiny one celled organism is actually a bacterium that can convert ethyl alcohol to acetic acid. A weak solution of acetic acid, usually about three percent, is commercially prepared as white vinegar. Dark brown vinegar is usually derived from the alcohol generated when apple cider is fermented. Other types of vinegar, such as wine vinegar, are also available commercially.

Tumors in Plants

Plants do indeed get cancer. One of the most common methods by which plant tissues are induced to form tumors is due to infection with the bacterial organism *Agrobacterium tumefaciens*. This organism is widely present in the soil. It is sometimes able to invade a plant through wounded tissues. *A. tumefaciens* is frequently called the "crown gall bacterium". The microbe can insert portions of its own genetic material (DNA) into healthy plant tissues and alter them to the point that excessive hormones are secreted. This shift in the plant's metabolic activity can cause tumors to begin to grow.

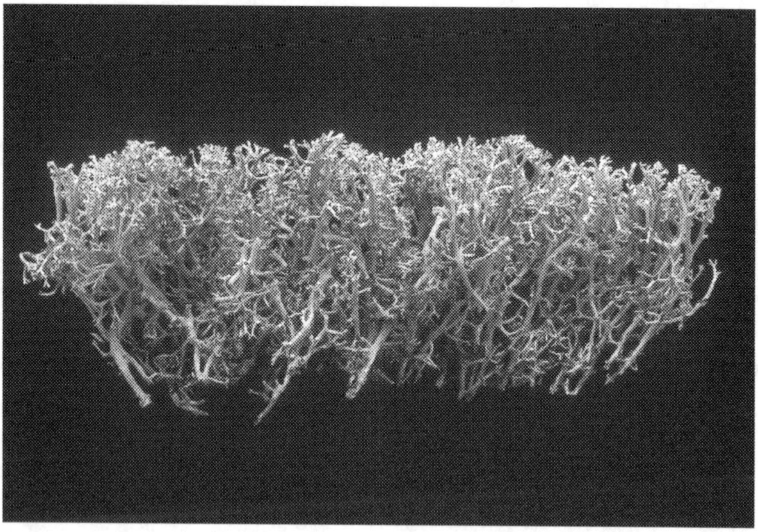

Close up of fruticose lichen. Courtesy of United States Department of Agriculture, Agricultural Research Service (USDA-ARS). Photograph by Peggy Greb, collected by Joanne Romagni.

Lichens: A Partnership

Lichens are familiar to almost everyone. They are, however, difficult to classify. This is due to the fact that lichens are actually a close association between two organisms. One member of the partnership is called the phycobiont because it is plant-like in habit. Phycobionts will either be a species of green algae or cyanobacteria. These organisms carry out photosynthesis to make food for the partnership. The other member of the association is called the mycobiont because it is a fungus. The mycobiont penetrates the phycobiont to absorb food but contributes water and minerals it absorbs from the environment. Both the mycobiont and phycobiont carry out cell division to make the lichen body or thallus (plural is thalli) increase in size. The growth is painfully slow. In some cases the lichen may increase in size at a rate as small as one hundredth of one millimeter per year. When pieces of the thallus break away, they may grow to form a new lichen. Some lichens prepare and release specialized rounded masses of thallus called soredia that are used for asexual reproduction. Lichens are commonly found growing on trees, rocks and other habitats. Very tight, crust-like thalli are called crustose lichens. Squamulose lichens have scale-like thalli. Lichens with leafy thalli are called foliose and those with slender, branching threads are known as fruticose lichens.

Algae Has Many Uses

The algae (singular is alga) include a wide variety of species that range from microscopic, single celled organisms to giant sea weeds that may reach more than one hundred feet in length. Many biologists classify the algae in the plant kingdom while others place them in kingdom Protista. Some species of algae are edible and are even grown commercially. Many red and brown algae produce a chemical called algin in abundance. It is a gel-like material used to make medicine capsules and to provide stability for various sorts of cosmetics. Algin and similar chemicals are also used to make agar, the basis for culture media in microbiology. Ice cream and paint may also include these chemicals to give them long-term stability and prevent separation of ingredients. Several species of red and brown algae are unusually rich in iodine. In the past they were commercially harvested and the iodine was extracted for medicinal uses. Algae have been used to fertilize the soil, and as food for cattle.

Giant Puffballs

Have you ever seen a fungus the size of a beach ball that emits hundreds of billions of spores? If you live in the United States (particularly in the eastern parts) you may be familiar with *Calvatia gigantea,* the giant puffball. Populations are also known in Europe. These organisms are common in pastures and similar habitats. The smaller ones may only reach the size of a ping pong ball. Some of the larger ones weigh more than 15 pounds. Giant puffballs begin as a white-colored sphere. As they mature they change color to tan, then brown. With time, they dry and crack open to release their massive store of spores.

Bacteria That Varnish the Desert

In many desert and desert-like environments around the world some rocks are covered by thin layers of pigments that may range from orange to red, and brown to black in color. These pigment layers have come to be known as "desert varnish." Amazingly enough, desert varnish is the product of several different species of hard working bacteria that grow upon the rocks. The colorant material is secreted by the bacteria as a byproduct of their metabolic activities. Desert varnish may also help to protect some of the bacteria from environmental extremes. The material is primarily composed of tiny clay grains and chemical compounds containing iron and manganese. In a few places samples of ancient artwork, inscribed by Native Americans within the thin layers of desert varnish, survive.

Bread Mold

Several species of fungi may grow upon bread if the environmental conditions are right. One of the most common is *Rhizopus nigricans*, black bread mold. Spores of the fungus are found almost everywhere. If they germinate, they will quickly give rise to a mass of fungal filaments. In time, black spore cases may be observed among the growth. The fungus reproduces sexually and excretes a material called "fumaric acid" that has many uses in industry.

Cells in Glass

Types of algae known as diatoms have a glassy-cell wall that tends to endure long after the organism itself dies. The cell wall actually is glass-based, being rich in silica. The cell walls of diatoms work a lot like a box and lid. The two halves are known as valves. One valve is larger than the other and fits upon it. These glassy remains form diatomaceous earth which is used in some types of polish, toothpaste, filters and artwork. Diatomaceous earth is also an environmentally friendly insecticide. The tiny, sharp glassy fragments found in diatomaceous earth can actually puncture the exoskeletons of some insects and leave them vulnerable to drying out. Living diatoms store excess food material as oils.

Pseudomonas, the Janitor

Some bacterial species in the genus *Pseudomonas* possess the unusual ability to utilize unrefined oil (crude oil) for an energy source. They make enzymes and other chemicals that act almost like detergents on the oil. These bacteria have been used directly to clean up residual oil on oil tankers. They have also been used to degrade oil spills. Scientists continue to study the oil-eating *Pseudomonas* species and have started to genetically engineer their oil-attacking cellular products.

Yeast

These organisms have long been used to make beer. Two species in particular, *Saccharomyces cerevisae* and *S. carlsbergensis* are widely used today in the brewing industry. As the organisms ferment or break down sugars in grain (usually barley) they release alcohol and carbon dioxide as byproducts. If the yeasts complete most of their work on top of the grain mixture, ale is the product. Bottom fermentation produces lager. Flowers from hops or other plants may be added to the mixture for flavor. The process of wine-making is similar except that fruits such as grapes are used as a sugar source instead of grain. Yeasts also make bread, pastries and other baked goods rise as the carbon dioxide gas they produce is trapped within the dough. The alcohol evaporates away during baking.

They'll Eat Anything

Some species of bacteria are known for being able to utilize strange chemical compounds in their diets and metabolic processes. A few examples of unusual substances known to provide raw materials for bacterial metabolism include iron, oil, arsenic, cyanide and gold. Biologists eagerly research these organisms, not only to learn more about their unusual metabolic activities but also to find potential industrial uses for them.

How Very Icky

Aquarium-keepers are, unfortunately, usually well acquainted with the fish disease known as ich (sometimes spelled ick). This infection produces patches of white growth on aquarium fish (as well as on wild freshwater fish) and can cause death. A whole tank of fishes may succumb within days. The culprit is a protozoan in the phylum Ciliphora, *Icthyophthirius multifiliis.* One stage in the life cycle of this organism takes place at the bottom of the aquarium. Free swimming protists are produced in great numbers during this time. They are able to enter the skin of fish to cause the disease known as ich.

Jelly Fungi

Some fungi that live within dead wood produce their reproductive structures in the form of small blobs of jelly-like material. The bulk of the organism is actually found within the dead woody tissues where it feeds. These curious organisms belong mostly to the orders Dacrymycetales or Tremellales. They are commonly called "jelly fungi." Depending on the species, the jelly-like mass may be clear to white, yellow or orange in color. It produces and releases spores that spread the organism to other locales within the environment.

The Fire Algae

Several species of algae are placed in the division Pyrrhophyta and are known by monikers such as "fire algae" and "dinoglagellate." Many species are bioluminescent; they emit light as a byproduct of energy production. The name dinoflagellate is derived from the fact that the organisms typically have two whip-like flagella that they use for locomotion. Both freshwater and marine species exist. Some are colonial, others unicellular. Quite a few species live within animals, such as jellyfish and coral, as part of symbiotic relationships. Some species of pyrrhophytans cause an ecologically significant phenomenon known as red tide. Several organisms, all of which can produce a toxin that will interfere with the nervous tissues of various animals, may reproduce to form a "bloom" or high density population in a localized area. Various animal life in the area may die or, at the very least, become contaminated with the toxin. In some extreme cases even mammals as large as a dolphin have died as a result of contact with the toxins.

What's in a Name?

Throughout history, fungi were regarded as unusual organisms due to their lack of chlorophyll and association with damp, dark habitats. Some of the most interesting common names ever applied to any type of organisms were the result. The names conjure images of fairies, evil spirits and other supernatural things.

Examples include elf saddle, black knot fungus, dead man's fingers, fairy butter, witches' butter, stinkhorns, old man of the woods, witches' broom, chicken fat fungus, jack-o-lantern, destroying angel mushroom, death cups, fairy rings and pig's ears.

The Lowly Paramecium?

The common name paramecium is applied to several species of protists. These microscopic organisms are often thought of as primitive but they can do some amazing things. A single cell makes up the entire organism. In this unicellular organism is found some specialized cell parts, or organelles, that are akin to the body organs of higher animals. There is a contractile vacuole that acts much like a kidney to expel excess water from the cell. Paramecia also have a mouth-like structure called the oral groove. They form a cytoproct or anal pore for release of waste products from the cell. These incredible organisms make their own energy and digest their own food. They are capable of incredible types of movement due to the presence of beating, hair-like cilia on their cell surface. Paramecia can also defend themselves by releasing sticky threads, called trichocysts, which are embedded in their cell membranes.

Record Bacterium

The record for the longest known bacterium probably goes to *Epulopiscium fishelsoni.* This species was discovered living inside the abdominal cavity of the surgeonfish. Its length is about one fifth of one millimeter to almost three fourths of one millimeter, extraordinarily large for bacteria. In fact, most bacteria are barely visible with microscopic magnification of 1000 times. On average *E. fishelsoni* is about the size of a printed hyphen, like that shown at the end of this paragraph: -

Rusty Coffee

Today the British are noted tea drinkers. This was not always the case however. Prior to the mid 1800s, coffee was the drink of choice among most of the British. Tea was used sparingly. About this time a fungus, known as rust, ravaged through coffee plantations. Prices escalated and many of the British switched to drinking tea due to its cheaper cost and easier acquisition. Tea remained popular as a pleasure drink even after the price of coffee became affordable again.

Buy Stock in this Protist

Nosema locustae is a protozoan that has been sold as a biological control agent for grasshoppers, some cricket species and similar organisms. Spores of the organism may be added to food or sprayed in liquid form. When grasshoppers and related insects eat these spores, the protist becomes active within their digestive tract. In young insects death is quick. It may take longer in more mature individuals. Since many grasshoppers eat the dead of their own species, the spores of the lethal microbe may be spread in that fashion as well. *N. locustae* has a very specific host requirement and will not harm other organisms.

Milk Makes the Grade

Milk can harbor all sorts of microorganisms, including some deadly species of bacteria. Many of these organisms are picked up from the environment, after the milk is removed from the dairy animal. Cow's milk is often graded according to the approximate numbers of microbes it contains. Bacteria and other microorganisms even find refuge in pasteurized milk. Among the cleanest milk is certified milk. It is often used in the manufacture of cheese and ice cream. In one system of grading, certified milk contains fewer than 10,000 microbes per milliliter of milk. The most commonly seen commercial milk packages are Grade A. In one system of grading, a sample must contain fewer than 30,000 microorganisms per milliliter to earn this grade. Grade B is assigned when the number of microbes detected is between 30,000 and 50,000.

Fungi that Trap and Kill

A few species of fungi are adapted to actively trap and digest living organisms. Some have sticky chemicals in their thread-like hyphae. The hyphae penetrate the soil. Small animals and microscopic organisms may become trapped and ingested for food. Among the most amazing fungi are those that trap and digest small roundworms known as nematodes. One species, *Arthrobotrys dactyloides,* sets snare-like traps made of its hyphae in the soil. As the nematodes burrow into the snares, they tighten around the body of the worm like a noose.

Salty Algae

Some algae are able to thrive in salt water conditions that would cause most ocean-dwelling creatures to perish. *Dunaliella salina*, from Australia, is one such species. It can float upon and live within bodies of water that are saturated with salt. Another amazing species, a green alga known to biologists as *Dangeardinella saltitrix*, is so salt tolerant that it can actually live within solid salt crusts at the shores of lakes and other bodies of water. When actually in salt water, the organisms use flagella to swim about.

Keeping Nitrogen in Circulation

Nitrogen is one of the most abundant elements on the planet. Plants are unable to use nitrogen in its gaseous form but must utilize it in other ways. Monerans (bacteria and cyanobacteria) play a tremendous role in keeping nitrogen cycling through the Earth's ecosystems. Nitrogen fixing bacteria and cyanobacteria are able to convert gaseous nitrogen to a nitrogen-containing compound called ammonia. The organisms do this by combining the nitrogen gas with hydrogen gas in the presence of enzymes they make. Examples of monerans that can make the conversion include *Rhizobium*, which lives within the roots of many types of plants, and *Anabanea* among many others. Ammonia or its products may be further acted upon by various species of the bacterial genera *Nitrobacter* and *Nitrosomonas*. These species can convert the chemicals into simpler products (nitrates and nitrites) that plants can utilize directly from the soil.

Ancient fossilized stromatolites. Courtesy of Department of the Interior/United States Geological Survey (USGS). Photograph by Beatriz Ribeiro da Luz.

Stromatolites: Ancient Wonders

Some of the oldest known fossils are the stromatolites. The most ancient ones in fact are thought by some scientists to be more than three billion years old. Superficially they resemble nothing more than dome-shaped rocks and are associated with oceanic environments. Some are very tiny, others several feet tall and wide. The amazing thing about these rock formations is that they contain the fossilized remains of tiny monerans known as cyanobacteria. Very few active or living stromatolites remain on Earth today. Most are known from Australia. Biologists believe that a stromatolite begins with a floating, tangled mass of the microscopic cyanobacteria. Over time, the tiny organisms secrete lime, calcium carbonate, and other materials. Various types of debris from the environment also get trapped in the growing mass to help create the rocky formations.

Truly a Humongous Fungus

Fungi of the genus *Armillaria* are very common throughout the United States and Canada. Various species occur there and elsewhere in the world. Many species of this fungus feed upon dead and decaying trees. However, some can also attack and kill living trees by invading their root tissues. When this fungal disease appears on living trees it is sometimes called "armillaria root disease" or "shoestring root rot." The fungus produces black, stringy threads that sometimes reach up to one-sixteenth of an inch in diameter. A few especially large masses of fungi in the genus have been discovered in forests of the Western United States. Of course, most of the organism exists underground and within dead and dying trees. Their reproductive structures appear above ground in the form of mushrooms that erupt from the trees or nearby soil. Masses of *A. ostoyae* have been found in Washington and Oregon. Oregon is home to what is believed to be the largest fungal organism on Earth. It is said to cover more than 2,000 acres of forest land and is thought to be thousands of years old. Biologists have sampled the underground mycelia in various locales within the forests and have determined that they do indeed appear to belong to single, massive organisms. The reproductive structures of *A. ostoyae* are commonly called "honey mushrooms" and are said by some to be edible, but not particularly tasty.

One Durable Bacterium

Bacteria are known for their ability to survive in some very extreme environments. One species, *Deinococcus radiodurans*, must surely win the prize for being able to ride out the most extreme conditions imaginable on Earth. Not only is the bacterium highly resistant to dehydration and deprivation of nutrition, it can also survive extreme levels of ultraviolet and nuclear radiation. This organism was discovered in the 1950's and has been observed growing on food that was irradiated for preservation purposes. Levels of radiation are most often quantified in a unit called the rad. *D. radiodurans* has shown the ability to survive more than one and one-half millions rads. To put this in perspective, 1000 rads of radiation could easily kill a human. It has been reported that this incredible bacterial organism has even been found thriving in the cooling tanks of nuclear reactors.

Rust and Smut

Rust is the common name applied to about 4000 species of fungi that cause disease in some plants. The name has nothing to do with rusting metal. Throughout history rusts have been responsible for destruction of ornamental plants, crop loss and famine. There is evidence that ancient people of Rome viewed these destructive fungi as a curse from their god, Robigus. Rusts tend to have very complicated life cycles that usually involve more than one host organism. In many cases, the life cycle requires at least two years for completion. Examples of these destructive fungi include flax rust, bean rust, asparagus rust, coffee rust, white pine rust, cedar-apple rust, sunflower rust, wheat rust, gooseberry rust and many others. Smuts are related to rusts. There are more than 1000 species which cause disease in various plants. Their name derives from the black mass of spores by which the smuts reproduce. Examples include corn smut, onion smut and oat smut.

Mushrooms and Toadstools

Botanically speaking, mushrooms and toadstools are merely different common names for the same type of structure. Certain species of fungi use these structures for reproduction. The actual organism feeds and grows beneath the soil or within rotting wood or some similar habitat. The mushroom is a reproductive structure that contains the same thread-like hyphae making up underground portions of the fungus, but in a much more tightly packed form. Most mushrooms exist for only a few days or for several weeks. During this time they scatter spores into the environment for reproduction.

Red Algae

Some biologists call them plants, others call them protists. Still others insist that they are neither. However, the red algae (usually placed in the division Rhodophyta) remain one of the most interesting groups of organisms on the planet. All members of the group contain large amounts of a red pigment known as phycoerythrin which accounts for their usual red color. In some cases other pigments mask the red to produce organisms that are black or brown. Some red algae are unicellular but most are multicellular. They typically grow in marine environments attached to rocks or similar substrate materials. One particular group (in the family Corallinaceae) was mistaken for corals until the 1830s due to their thick, calcium carbonate rich cell walls. Some species of red algae are used to make nori, which are the wrappings of sushi rolls.

Bacterial Freezing

Microbiologists have isolated an enzyme from the bacterium *Pseudomonas syringae* that elevates the freezing point of water. In other words, in the presence of the enzyme, water will freeze at warmer temperatures. The bacterial enzyme has been used commercially in the preparation of snow at ski resorts. In nature the bacterium parasitizes a number of plant species including several types of trees, garden plants and grasses.

The Slime Molds

Slime mold is the common name given to a number of species of organisms that have migratory feeding stages in their complex life cycles. Some scientists classify the bulk of slime molds in the kingdom Fungi while others place them mostly in kingdom Protista. At any rate these organisms will form a traveling body known as a plasmodium, or a slug, depending on the species. The slugs actually resemble small worms or snails and could easily be mistaken for an animal as they travel in search of food. Plasmodia may be very small or up to several feet across. Some of the largest ones have been observed in yards, climbing over trees or other objects during feeding. After feeding is complete, the slime molds usually produce fruiting bodies that give rise to spores. The life cycle begins again at this point.

Giant Algae

Some species of algae can grow to lengths that are actually greater than the tallest trees. Organisms commonly known as giant kelp, in the genus *Macrocystis*, can achieve these great lengths. They may regularly grow at a rate of two feet per day. An account of a kelp plant just over 700 feet in length was once recorded. Dense underwater populations of kelp are sometimes called "kelp forests." Kelp and some of its smaller relatives are placed in the division Phaeophyta and may collectively be called the "brown algae." Many biologists consider these and other algae to be plants. One of the most economically important derivatives of brown algae (and some other algae types) is known as algin. It is extracted from the cell walls of these organisms and is used as a thickening agent in cake frosting, cosmetics and rubber. It has also been added to various foods such as ice cream and beverages such as beer.

Archaebacteria

Archaebacteria literally means "ancient bacteria." This term has been used to refer to a group of very primitive bacterial organisms that many scientists believe should be placed in their own biological kingdom. They have been traditionally classified with other bacteria and with the cyanobacteria in the kingdom Monera. Archaebacteria live in some highly unusual and exotic habitats. Some, such as members of the genus *Sulfolobus*, are found in boiling pools of sulfuric acid; others in water with high salt concentrations and still others in methane-rich bog-type environments. Microbiologists use the term halophile to describe bacteria that inhabit salt-rich environments. The genus *Halobacterium* is an example. Thermophile is a term used to describe those bacteria that thrive in high temperatures. Some may survive at temperatures exceeding 230 degrees Fahrenheit. The term psychrophile refers to those organisms that grow at very cold temperatures. Finally, the term methanogen describes those bacteria that produce methane as a byproduct while living deep in bogs and similar environments. Both freshwater and marine species are known. Genera include *Methanobacillus* and *Methanococcus*.

A Forest Altering Fungus

Along an extensive band, east of the Mississippi river, the American chestnut tree (*Castanea dentata)* was once an important plant of the forest. These trees originally ranged from Maine into parts of Florida. They attained their greatest size and numbers in the southern Appalachian Mountains. In fact it has been estimated that one of every four trees in those old forests was an American chestnut. These trees attained a great age and size. It was not uncommon for some to live well past 500 years. They were sometimes called the "redwoods of the east" due to their size. The nuts were an important food source for many organisms of the forest. Citizens of the Appalachian Mountains used the nuts for food as well. They also traded the nuts for food and supplies at community general stores. Chestnut wood was extremely resistant to decay and was used in a variety of ways including furniture, housing and as railroad cross ties. The trees were also very rich in tannin, which was crucial to the leather industry. Many towns in the Southeast owe their origin to the American chestnut tree. It supported the economy to such a great extent that some towns developed entirely because of the tree. In the early 1900s a fungal parasite was noted in the northern ranges of the tree. The fungus, *Cryphonectria parasitica*, is native to Asia and proved to be deadly to our native chestnut trees. However the fungus does not usually infect the roots of the trees, only the stems. In less than fifty years, up to four billion mature trees had been killed all along the tree's range. In many cases, the still-living roots of these giants send out short-lived shoots. Today the American chestnut tree is little more than a small understory tree of the forests. A small number of isolated older trees survive. Following the death of the American chestnuts, insect larvae invaded the wood to produce what is now known as wormy chestnut. The lumber is highly prized for building and crafts to this day. Efforts to bring the American chestnut tree back to its former place of prominence in the Appalachian forests continue.

Giant Ameba

"Giant" may, of course, be a relative term. The common name ameba is given to countless species of unicellular protozoans in the phylum Sarcomastigophora. They inhabit water, soil, and many other places. Some live as parasites within or upon other organisms. Amebas are generally microscopic and they move by means of cytoplasmic streaming and pseudopods. In essence, the organism extends its cell membrane into a foot-like projection. Its cytoplasm then flows into the foot. Therefore, the organism moves. Two species of ameba (*Chaos carolinense* and *Pelomyxa palustris*) are known to reach up to five millimeters across.

Magnetic Monerans

Several species of bacteria that synthesize and store a magnetic mineral known as magnetite (iron oxide) have been identified by biologists. Not only do these organisms make magnetite, they also appear to utilize it to help them find food and orient toward the magnetic poles of the Earth. This response is called magnetotaxis. Many species found above the equator will swim toward the North Pole. Those below the equator swim toward the South Pole. At the equator, the swimming response still appears to be directed in a polar direction; but some swim north and others south. Many animals have crystals of magnetite located in various body tissues. Scientists know very little about how it works in some species.

Fairy Rings

In yards and pastures, mushrooms are sometimes observed growing in a ring-like pattern surrounding an especially lush patch of grass. These formations are commonly known as "fairy rings" due to an ancient legend about fairies meeting there to dance. Fairy rings are almost always formed by small mushrooms in the genus *Amanita*. Below the surface of the ground the vegetative part of the fungus, the mycelium, grows in a circle. Its reproductive structures, called mushrooms, sprout along the ring of mycelium. As the fungus grows it releases important nitrogen-containing compounds into the soil. This accounts for the rich growth of grass commonly observed within a fairy ring. It is estimated that some of the larger fairy rings may be more than 500 years old. They remain unnoticed until mushrooms appear above the ground. The mushrooms are the reproductive structures of these organisms.

Euglena

Euglenas are remarkable microscopic protists in that most species possess both animal-like and plant-like characteristics. They move in a variety of ways but most often use a whip-like appendage, known as a flagellum, as their primary means of locomotion. Some species have more than one flagellum. Most euglenas live in fresh water and have a pump-like vacuole to rid their single cell of excess water. They have a definite head end and tail end. Most species even have chloroplasts that enable them to made food from the energy of the sun and from other raw materials. Many of these species even possess a light-sensitive structure known as an "eye spot." They seek out and swim toward areas of bright light.

Blue Green Algae?

A group of organisms formerly referred to as blue-green algae are worthy of consideration. A more descriptive term, cyanobacteria, is widely used today because the organisms in question are neither blue, nor green, nor algae. In fact, cyanobacteria belong to the same kingdom as bacteria (kingdom Monera). Their cells are prokaryotic, meaning that they have no true nucleus or specialized cell parts. Cyanobacteria may exist as single cells or as colonies. They are found in water or in moist land habitats. Like green plants, they carry out photosynthesis to make their own food. Some colonial cyanobacteria include *Nostoc, Rivularia, Oscillatoria* and *Anabanea* which tend to occur in elongated filaments made of individual bead-like cells.

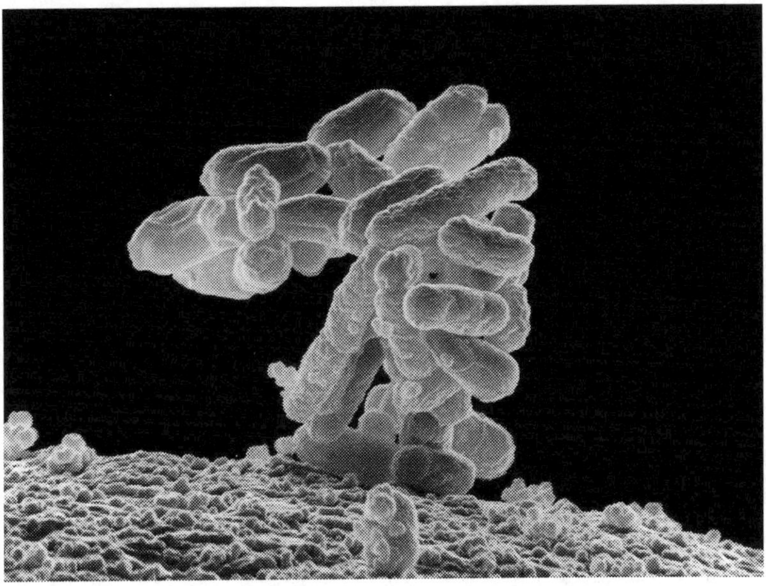

Electron micrograph of elongated *Escherichia coli* bacteria.
Courtesy of USDA-ARS. Photograph by Eric Erbe.

Unusual Cell Shapes

Some species of bacteria do not display the traditional round, elongated or spiral shapes with which so many people are familiar. Some species are completely without a rigid cell wall and have no consistent shape. The term pleiomorphic has been applied to such monerans. Square-shaped bacterial cells are known from species that inhabit the Red Sea. Triangular shapes may be found among bacterial organisms that inhabit hot, salty springs in certain areas of Japan. Star-shaped species of bacteria are known as well.

Accidental Fungicide

According to a fairly well documented legend, the first fungicide (an agent that kills or slows the growth of fungi) was discovered by accident in the grape vineyards of France in the 1860s. It seems that a fungus known as "downy mildew of grapes" *Plasmopara vitacola*, entered France from the United States about this time. The fungus was causing a great deal of concern for the wine industry. Vineyard keepers also had the occasional problem of having their prized grapes stolen and eaten by passersby. Some farmers found that an approximately 1:1 mixture of lime and copper sulfate applied to the developing grapes would temporarily make them so foul tasting that no one would dare steal them. The treatment produced no permanent alteration in taste or quality of the grapes. As luck would have it, this attempt to discourage grape thievery also proved useful in another way. It was observed that the lime and copper sulfate mixture killed the grape mildew. Thus, the first man-made fungicide was born.

Foraminifera and Radiolaria

Foraminifera are a group of marine protists that secrete and live within small, protective shells or tests made of calcium salts. The shells are usually microscopic but reveal beautiful and intricate shapes upon magnification. A few species produce shells that exceed microscopic size. Inhabitants of the shells are unicellular organisms related closely to ameba. They extend their cell membranes through tiny pores in their shells for purposes of movement. Some species live at depths of more than 15,000 feet. Organisms in the order Radiolaria are similar to the foraminiferans. However, their shells are mostly silica based.

Where Do the Holes in Swiss cheese Come From?

Bacteria of the genus *Propionibacterium* are mostly responsible for the formation of the characteristic holes in Swiss cheese. These organisms generate carbon dioxide as a byproduct of their growth within the developing cheese medium. The carbon dioxide gas forms large bubbles that rupture, leaving behind holes as the gas escapes. If a block of cheese has no holes, it is said to be blind. This is because the holes are sometimes referred to as eyes. Baby Swiss cheese tends to have much smaller holes than the average product.

Bewitching Fungus?

Claviceps purpurea is a fungus commonly known as ergot. It parasitizes a number of grains, including rye. During cool weather tissues of the fungus develop into a black mat on the host plant. These tissues contain lysergic acid diethylamide, otherwise known as LSD, as well as similar chemicals. In sufficient quantities these substances can cause hallucinations, pain, depression, headache, unusual skin sensations, muscle spasms, convulsions and a number of other unusual symptoms in humans and other mammals. Some also cause a powerful constriction of blood vessels, producing a sensation of burning pain in the extremities sometimes known as "Saint Anthony's fire." In the mid 1970s a psychologist and researcher presented compelling evidence that an outbreak of ergot poisoning may have been responsible for the witchcraft hysteria of Salem, Massachusetts in the late 1600's.

Fungal Parasite Names

Some of the most disturbingly descriptive common names are assigned to fungal parasites of plants. Examples include cabbage club root disease, apple scab fungus, tar spot of maple, plum fruit pocket disease, brown rot of peach, cedar-apple rust, apple bitter rot, apple blotch, black pox of apple, white pine blister rust, oak leaf wilt fungus, powdery mildew, corn smut and potato powdery scab.

A Sea of Algae

One of the most unusual ecosystems on the planet is the Sargasso Sea. It is found within the Atlantic Ocean, off the eastern coast of North America, north of South America and west of Africa. The environmental conditions for the Sargasso Sea are created in part by a set of oceanic currents that combine to form a clockwise course of flow. The major currents include the Gulf Stream and the North Equatorial Current. The boundaries of the Sargasso Sea shift and change with the water currents. It tends to maintain an approximate area of two million square miles. The water within is much more calm and warm than the surrounding ocean and it tends to be very deep blue in color. The sea derives its name from the extraordinarily large concentrations of *Sargassum* algae that tend to drift there and become trapped within the circular currents. Although the algae do not cover the entire surface of the Sargasso Sea, they do collect to form great tangled mats. *S. natans* is the most common species found but a few others do exist. All of the *Sargassum* is free-floating due to the presence of air bladders or pneumatocysts on the plants. These air bladders superficially resemble brown grapes. In fact, the name is derived from "sargaco," the Portuguese word for grape. It is said that Christopher Columbus was the first person to formally describe the Sargasso Sea. Many legends about ships becoming tangled within the mats of *Sargassum* have been retold through the centuries. However, it appears to be highly unlikely that the algae could actually be strong enough, or be present in great enough concentrations, to entangle a large ship. Some biologists referred to the Sargasso Sea as an oceanic desert in the past. While it does appear that very little life exists within its depths, there is a complex community of organisms to be found among and above the seaweed mats. Some highly adapted species of fish and crabs are found there with various species of turtles, octopuses and birds, just to name a few. Many types of microscopic life also make the Sargasso Sea their home.

Fungi and Cheese

Many, many types of cheese are produced throughout the world. Both bacteria and fungi are involved in cheese manufacture. Some species of fungi impart particular characteristic flavors of certain cheese varieties. Fungi of the genus *Penicillium* flavor Camembert and Roquefort cheeses for example.

Sexual Reproduction among Bacteria

Who would believe that bacteria, organisms that are regarded as very primitive, would actually have methods in place by which the genetic material of DNA could be mixed between individual organisms? Some species are able to do just that and there are a variety of means by which it may happen. Some species carry out a form of sexual reproduction known as conjugation. The individual unicellular organisms actually join by a pipe-like connection known as a pilus so that genetic information, in the form of a plasmid, may be passed from one organism to another. Bacterial transformation may also allow the mixing of DNA between organisms as the genes from a dead bacterium may be incorporated into a living bacterial cell. Viruses may inadvertently carry DNA from one bacterium to another in a process known as transduction.

A Champion Reproducer

The protist organisms in the genus *Glaucoma* are noted for their rapid rates of reproduction. Some species may divide as often as once every four hours when environmental conditions are very favorable. Such rates are among the fastest known among these types of microbes.

A Fungus That Changed History

Phytophthora infestans is a fungal pathogen of Irish potatoes, producing a disease called "late blight." The cells of the fungus enter the leaf and stem tissues of the potato plant and begin digesting the host alive. Farmers continue to battle this parasite even today. In the mid 1800s the potato blight fungus produced a devastating chain of events that altered history. Ireland had a population of nearly 10 million people in the early 1840s. The potato was a staple of the diet at this time. A major outbreak of the fungus in the mid-1840s was responsible for more than one million deaths due to starvation. Many citizens of Ireland left their country and migrated to the United States during this time. Today the population of Ireland is about one-half that of the 1840s, before the infamous Irish potato famine changed the course of history.

Bdellovibrio

The genus of bacteria known as *Bdellovibrio* includes species that actually prey upon certain other bacteria. They are fast-moving and strike their prey bacteria by entering their cell walls. *Bdellovibrio* digests its prey while nestled between its cell membrane and cell wall, an area known as the periplasmic space. Next the *Bdellovibrio* undergoes rapid cell division, causing the host cell to rupture and thereby release more predatory bacteria.

Metaparsatism

It may sound like the old children's song about the lady who swallowed the fly. Parasites can have parasites. It is true! A good example is the protozoan genus *Entamoeba.* It contains some parasitic species that live within another protist of the genus *Opalina*. That genus is known to parasitize certain types of frogs.

When You Eat Chinese, Thank a Fungus

If you enjoy soy sauce with your Asian cuisine, the fungus *Aspergillus oryzale* deserves your thanks. Soy sauce is a byproduct of the fungi's action on a soy bean based precursor material. The product derived from *A. oryzale's* work is called koji. Koji is mixed with salt and further acted upon by yeasts and bacteria to produce soy sauce.

Bacterial Insecticides

Various genetic strains of the bacterial organism, *Bacillus thuringiensis*, are being used in the United States as insecticides. The organism lives and reproduces within the digestive tracts of various insect larvae. Host organisms as diverse as moth larvae and mosquito larvae are killed by a toxic material released by the bacteria. Use of the toxin commercially goes back at least as early as the 1920s. Most regard the material as causing no harm to the environment.

Cold Algae

More than 300 species of algae are collectively referred to as the snow algae. They thrive upon and within snow drifts, glaciers and even icebergs. These algae contain pigments that may impart blue, gray, orange, green, red and pink colors to the snow and ice around them. Two of the most interesting species are *Chlamydomonas nivalis* and *Chloromonas rubroleosa.* Both are classified within the green algae but contain an abundance of red pigment. The colorful streaks and patterns made by the snow algae were a great mystery for many years.

Glow in the Dark Bacteria

Bioluminescence is known in all kingdoms of organisms. Some bacteria may produce light as a byproduct of their metabolic activities. A particularly noteworthy species is *Photobacterium phosphoreum*. This bacterium lives in marine waters and may be found in abundance upon the skin and internal organs of various animals including fish, lobsters and clams. They cause their hosts to emit an eerie glow.

Dutch Elm Disease

A fungal organism known as the Dutch Elm fungus, *Ophiostoma novo ulmi*, is often spread to healthy elm trees by way of a beetle. Once established on the tree's bark, the fungus enters the living tissues of the tree to feed. It also produces toxins that hasten the host tree's death. The first severe outbreak on record was in the United Kingdom in the 1920s. Millions of elm trees died but the population slowly rebounded. A new strain of the fungus appeared by 1975, killing more than 20 million trees. The fungus entered the United States in the 1930s destroying many of our native elm trees, *Ulmus americana*. The pathogen is still a problem today. Efforts continue to keep the killer under control.

Alcoholic Drinks

Bacteria and even some fungi are responsible for the alcoholic content of many drinks. As they make food from the sugars stored in various plants, they release alcohol as a byproduct of their fermentation activities. Humans learned early on to culture these organisms in various sugar sources and distill the resulting liquid to drink. Vodka comes from fermentation of potatoes. Tequila comes from cacti and sake from rice. Wines are derived from the fermentation of many types of fruits including grapes, apples, blackberries and peaches. Beer and whisky are made from various grains including corn, rye and many others.

Giant Cells

Green algae in the genus *Derbesia* are noted for having some of the largest cells known outside of the animal kingdom. Specialized cells form in some members of the genus that may reach one inch across. Most plant and protist cells are visible only with a microscope.

CHAPTER TWO

KINGDOM PLANTAE

Those Things Are As Big As Trees

Tree ferns are real and they grow large. Several species of ferns have earned the common name. Some may grow up to 30 feet tall and have fronds that are in excess of eight feet in length. Many others are small enough to be grown in pots but still maintain a growth pattern that is tree-like. Genera that include tree ferns are *Cibotium, Dicksonia, Cyathea* and *Nephela*, among others. Tree ferns are mostly limited to rainforest habitats and may be found growing wild in Australia, New Zealand, Hawaii and New Guinea. Several species are grown commercially as ornamental plants.

Fallen Leaves

Every tree periodically looses at least some of its leaves. Leaves may be lost due to age, injury or seasonal changes. At the base of each leaf is a region of tissue known as the "abscission zone." When the leaf's useful life has passed, the cells in the abscission zone begin to die. Chemicals called pectin, which help hold many leaves in place, begin to break down at this time as well. The dying leaf may turn brown or some other color, based on the type of pigmentation it has. In many trees brilliant colors show in autumn as the green chlorophyll begins to break down. The dead leaf is usually held by a fragile strand of xylem, analogous to a vein, until it is broken away by wind or rain. A plant that regularly looses all of its leaves each year is known as a deciduous plant. Maples, oaks and elm trees are examples. Those that loose only part of their leaves at one time are called evergreen plants.

Examples include the American holly, southern magnolia and most pine trees.

Christmas Cactus?

Many people enjoy growing these beautiful houseplants. Some have survived for decades to become family heirlooms. The foliage is deep green and the blooms may be white, pink, red, purple or orange. Despite the common name, these plants are not cacti from some exotic desert but are actually native to the Americas. They are tropical plants. Two types are commonly grown as houseplants. The Christmas cactus, *Schlumbergera truncata*, blooms in fall. The Easter cactus, *Rhipsalidopsis gaertneri*, blooms in early spring. Both plants must receive extended exposure to darkness a few weeks before the flowering season to insure a beautiful crop of blooms.

Annuals, Biennials and Perennials

The life cycles of most plants can be easily categorized according to the three terms presented above. Those plants that live for only one year or less are called annuals. They mature and produce seeds during this time. Corn, annual rye grass and many types of wildflowers are examples. Some plants grow and store food during the first year and produce seeds the second year. These plants, like carrots, are called biennials. Perennial plants live for more than two years and often for decades. In some species the above-ground tissues die back during winter but the roots live on. In other species, such as trees, the plant continues to grow above and below ground. These plants may produce seeds for hundreds of years.

Why Houseplants Wilt

Most plants have rigid cell walls that surround the membranes of each of their cells. The cell wall is made mostly of carbohydrate and helps to support the weight of the plant. The other key element of support is water. When plant cells are full of water, turgor pressure is the result. As soon as the water level in plant cells begins to fall, the turgidity drops as well. This causes wilting, as cell walls are generally not strong enough to support plants without additional pressure from water within.

What is a seed?

Higher plants utilize seeds for the dispersal of their young. Seeds actually contain a plant embryo, in a dormant state, as well as stored food known as endosperm. Seed plants produce the embryo by joining a female reproductive cell (known as an ovum) with a male reproductive cell, called a sperm cell. The sperm cells are carried about within pollen grains. When the ovum and sperm join, the resulting cell is called a zygote. The embryo develops from it. Non-seed plants like ferns, mosses and algae also use sperm and ova for reproduction. However, they do not produce embryos enclosed in a seed.

Fern Anatomy

Ferns and their relatives, such as horsetails and club mosses, belong to a group of plants known as the pteridophytes. Some biologists call this group Division Pteridophyta. Like other members of the group, ferns reproduce by means of spores. They are more advanced than the bryophytes, such as moss, in that they have vessel-like conducting tissues for food and water. Many genera and species have been described. However, some typical fern anatomy may be discussed. Most ferns possess an underground stem, known as a rhizome, from which roots and leaves arise. Developing fern leaves of many species become visible as highly coiled structures known as "fiddle heads" or "shepherd's crooks." As the leaves mature, they uncurl and continue to grow. Mature fern leaves are usually called fronds. The tiny leaflets on a larger leaf are known as pinnae. Like so many other plants, ferns lead a double life when it comes to their reproductive cycle. They do not produce flowers but do reproduce sexually. Ferns typically follow a pattern of alternating growth phases that will produce either spores or gametes. The image most of us call to mind when we hear the word fern is that of the sporophyte generation that makes spores. Some spores are formed in clusters called sori that occur on the underside or edges of some species of ferns. In other species, the spores may be produced on specialized fertile fronds. The spores give rise to the gametophyte generation that produces ova (egg cells) and sperm. Once fertilization occurs, the first young fern frond of the newly produced sporophyte generation emerges from a structure known as the prothallus. The prothalli of many ferns are visible without magnification but are usually inconspicuous and go unnoticed by the average person.

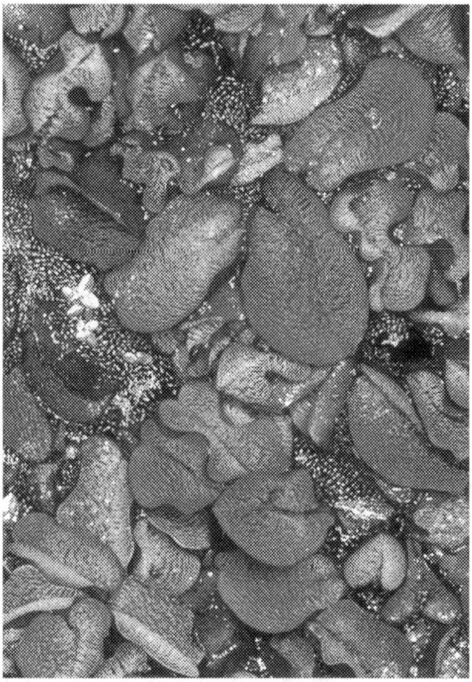

Close up of floating fern *Salvinia*. Smaller plants are duckweed. Courtesy of USDA-ARS. Photograph by Scott Bauer.

Ferns That Float

Ferns grow in the woods or in pots, right? Well, not all of them. Two families of ferns, Azollaceae and Salviniaceae, feature one genus each of strictly aquatic, floating organisms. The plants' leaves float on the surfaces of ponds, slow streams and similar habitats while their roots hang below the water surface. Members of the genus *Azolla* are often called by such common names as mosquito fern, red azolla and red water fern. The leaves of most species are less than two inches long. However those of *A. nilotica*, a species found in Africa, may sometimes exceed one foot in length. Under ideal conditions, which include abundant shade, *Azolla* ferns grow in lush green mats. In conditions of intense light, their leaves fade to a brown or red color. Members of the genus *Salvinia* are often called "water spangles." They superficially resemble floating clover leaves. Both genera are valuable parts of the ecosystems in which they occur. Fish and other animals may eat the plants. Also, a thick surface covering of ferns may discourage growth of submerged weedy plants. Both genera have been widely grown in ponds and aquaria around the world. Some farmers have added species of *Azolla* to their rice fields to help enhance growing conditions.

A Real Corker

Almost all trees produce a material known as cork from specialized tissue called the cork cambium. At maturity, cork cells are non-living and contain water-proofing materials. Cork tissues help to protect the more delicate living cells beneath the bark of trees. In past years, cork was harvested to make bottle stoppers. Today cork is used in insulation, gaskets and other commercial products. Some trees make very little cork while others are preferred for commercial cork harvesting due to the large amounts they produce. The best known and most widely harvested cork-producer is *Quercus suber*, commonly known as the cork oak.

Many Forms of Pepper

Pepper is applied as a common name to many, many plants and plant products. Sorting out the classification of various common edible pepper types may be difficult. Many of the forms of pepper used as a shaken or ground spice come from a single species of plant, *Piper nigrum*. This woody, climbing, flowering plant of the family Piperaceae is native to India but is in cultivation in several locations today. After flowering, the plants produce clusters of fruits known as peppercorns. They first appear as green, berry-like structures and later ripen to a rich red or brown. The harvest time and treatment of these fruits determines the taste and color of ground pepper products. Young, unripe berries are quickly harvested and dried. They are ground to produce green pepper. This spice is not to be confused with bell pepper, a fruit, discussed later. Green fruits that are allowed to ferment will yield the familiar black pepper, widely used in pepper shakers, after they are dried and crushed. Another variation in the technique will produce white pepper. In this case, ripened fruits are stripped of their red coverings before processing. Finally, if the fruits are allowed to fully ripen, the whole structure (with the red skin intact) may be ground to produce red pepper. Most of the plants with pod-like fruits, that we call peppers, belong to the family Solanaceae. The primary genus for peppers is *Capsicum*. They generally grow as small, herbaceous plants. Depending on the degree of ripeness, and the species (or in many cases the variety or breed), the fruits may be called by such names as chili pepper, cayenne pepper, sweet pepper or bell pepper.

A Woodland Christmas Gift

One of the most common plants of the Eastern United States forests is known as the Christmas fern. They are native to the area and relatively common in most any suitable habitat, especially near streams. Christmas ferns have dark green, thick, leather-like leaves that botanists call fronds. The fronds grow from a woody underground stem. In very healthy plants, Christmas fern fronds may reach heights of more than two feet. Although individual fronds regularly turn brown as they die and collapse to the ground, the Christmas fern remains green throughout the year. These plants are one of the few green things that can be found in the wintertime forests. Because of this, it is said, Christmas ferns were often used as Christmas decorations by early Americans. Some have claimed that the leaflets remind them of tiny Christmas stockings as well. Biologists know the Christmas fern by its scientific name, *Polystichum acrostichoides.*

How Plants Grow

Most plants achieve growth in length and diameter due to a group of cells that are collectively known as "meristematic tissues." These cells reproduce rapidly. Near the tips of roots and stems, the tissues are referred to as "apical meristems." They cause primary growth or an increase in length. A person could carve their initials in a tree trunk and return decades later to find that the carving was still at the same height, relative to the ground. In other words, trees grow taller from their tops, not from the trunk. Secondary growth is the term applied to growth in diameter. The tissue responsible is called the vascular cambium and is found just beneath the bark of woody stems. Grasses lack vascular cambium and bark-producing tissues. They achieve their limited secondary growth by way of tissues known as the "intercalary meristem."

Where Do Seedless Fruits Come From?

Many fruits like bananas, grapes and pineapples are commonly sold in seedless varieties. How can this be? Where do the fruits come from if they have no seeds? In all cases, the plants are grown from asexually propagated plants that contain a natural or artificially induced mutation preventing the development of seeds. Bananas were among the first plants to be cultivated on a large scale in this fashion. Rare plants without seeds were collected from the wild and grown on plantations because people found the seedless fruits to be more appealing.

Norfolk Island Pine Trees

These huge trees, *Araucaria heterophylla*, are native to the Norfolk Islands near Australia. Some people refer to them by another common name, star pine. Each growing season, Norfolk Island pines increase their height by growing upward and adding a new shelf-like set of evergreen branches. In their native habitats some trees may reach impressive heights of more than 65 feet. These are the same trees that are so often grown as houseplants in the United States. Each year, especially around Christmas, hundreds of thousands are sold.

You Just Think it's a Moss

Mosses produce no flowers, seeds or fruit and have no specialized conducting tissues (xylem and phloem) for transport of water and food. The term moss is used erroneously in many common names of other organisms. Club moss, for example, is the common name of a group of plants that are more closely related to the ferns than to mosses. They are classified within the genera *Selaginella* and *Lycopodium*. A species of *Selaginella* is sometimes called "spike moss." The common name "moss ball" is sometimes applied to colonies of the algae *Aegagropila sauteri*. Irish moss is actually a red alga and pixie moss is a flowering plant. One of the most widely misunderstood plants is commonly called Spanish moss. It is often associated with the Deep South where it grows in dense mats among massive tree branches. This organism, in the genus *Tillandsia*, is actually a flowering plant and a member of the pineapple family. It uses the tree branches for support but makes its own food.

Largest Flower

The plant species with the largest flower known to science is found in Indonesia. The plant's scientific name is *Rafflesia arnoldi*. Its mature flowers, consisting of five orange-brown petals with white spots, may measure up to three feet across. The flower emits a disagreeable smell that lures pollinators such as flies and beetles. The scent is said to resemble that of rotting meat. Therefore a common name for the plant is "corpse flower." Interestingly enough, the very rare plant that sports this massive flower is parasitic on another plant species. The vegetative body of R. *arnoldi* is very inconspicuous and lies mostly within its host. It is only when the flower begins to develop that the plant attracts much attention.

Plants with a Sense of Touch

Many plants are indeed able to sense and respond to touch. A growth response to touch is known as a thigmotropic response. A movement in response to touch is called a thigmotaxic response. Several examples in the plant kingdom exist. *Mimosa pudica*, known by some as the "sensitive plant," grows wild in many parts of the United States. It is sometimes grown as a houseplant. When touched, it folds and drops its leaves. Some botanists believe this is an adaptation that helps the plant escape being eaten by grazing animals. Some carnivorous plants that feed on insects and other small animals also respond to touch. The modified trap-like leaves on the Venus fly trap snap shut when two sensory hairs are touched simultaneously. Likewise, the sundew bends its sticky hairs downward to entangle and trap an insect. Plants such as vines and roses adjust their growth responses in order to climb toward sunlight.

Plants with Quills and Big Sacs of Spores

Most botanists have seen but one genus in the family Isoetaceae, that being *Isoetes.* The most widely used common name for this genus is quillwort. The average person may have seen a quillwort but, chances are, misidentified it as a clump of grass. That description is only superficially accurate. These plants grow in or near ponds, streams and swamps. They range from a few inches in height to about one foot. Quillworts are close relatives of ferns. They produce their spores in a sac-like case known as a sporangium. As a group, quillworts have the largest sporangia of any known group of plants. A second genus of quillworts, *Stylites,* was identified in the Andes Mountains of South America several years ago. They are believed to be unusually rare plants.

What Are Pollen Grains?

Pollen grains are used by plants that reproduce sexually to house and transfer sperm, the male gamete or reproductive cell. Most plants produce nearly microscopic pollen grains although many are large enough to be easily seen without magnification. Plants such as grasses and pine trees generally utilize wind to transfer pollen. Many decorative flowering plants like roses and lilies utilize insects, small birds, bats and other small mammals to transfer their pollen. Some plants require very specific organisms to pollinate their flowers. When the pollen grain is transferred to the female reproductive parts of a flower it sprouts a long canal called the pollen tube. This pollen tube grows deep into the female parts of the flower. Sperm cells within the pollen grain move down the tube to fertilize the female ova or egg cells.

What is a Vegetable?

From a botanist's point of view vegetables are not formally recognized. The things we call vegetables are actually various specialized plant organs. Spinach, for example, is a leaf. We commonly eat other leaves including lettuce, collards and cabbage. We season other dishes with the leaves of various plants such as bay, tarragon, sage and basil. Celery is an elongated stem called a petiole or leaf stalk. Asparagus is actually a plant stem. The sweet potato, turnip, carrot, parsnip, beet and radish are roots. Irish potatoes are specialized underground stems known as tubers. Onions and garlic are specialized stems known as bulbs. The Hawaiian food known as poi comes from the taro plant. It is specifically derived from a modified stem, known as a corm. A tomato is actually (botanically speaking) a fruit. It is a ripened ovary containing seeds. In the 1890's however the supreme court of the United States rendered a decision that tomatoes are vegetables, not fruit.

Largest Seed

The record for the plant with the largest known seed would go the coconut palm tree, *Cocus nucifera*. These trees may grow up to 90 feet tall. The seeds are produced each year on mature, healthy trees. We commonly refer to the seeds as coconuts. They develop singly, within a large nut. Many people refer to the fibrous tissues surrounding the whole coconut seed as a husk. The husk fibers are known as coir in many parts of the world. These tissues are almost always removed from the coconut seed before shipping to commercial markets. Coir is often utilized in many non-food products as well.

These Ferns Hang Around

One of the most unusual growth patterns known among the ferns is that of *Ophioglossum pendulum*, otherwise known as the ribbon fern. In order to fully understand this plant's habitat, a short consideration of those plants known as epiphytes is necessary. Epiphytic plants grow upon the surfaces of other plants, usually trees. They are not truly parasitic but only utilize their hosts for space. They make their own food by photosynthesis and draw their needed water from the air or from a rain shower. Epiphytic plants are most common in rain forests and similar humid environments. The ribbon fern grows from a tiny spore that lodges within the root masses of other epiphytic plants. As the fern matures, its own roots mingle within those of the other epiphytes. It sends down extended, dark colored fronds that resemble long green ribbons. Some of the individual fronds may exceed four feet in length. The species is known mostly from Asia and Madagascar. It is found in some other nearby locales.

Flowering: The Long and the Short of It

The term photoperiod refers to the length of daylight in a 24 hour day. In many parts of the world, day length grows shorter during fall and winter and reaches its maximum during summer. Photoperiod has a dramatic impact on the flowering habits of many plants. The familiar poinsettia, for example, is known as a short-day plant. It requires extended periods of short-days in order for flowering to occur. Other examples of short-day plants include dahlias and soybeans. Daffodils, iris, corn and gladiolus are long-day plants. The ever-lengthening photoperiod of early spring is required to induce their flowering. Some plants are much less sensitive to day length and will flower at almost any time of year, provided they have some warm weather and sunlight. The sunflower, dandelion and rose are example of these neutral-day plants.

Plants with Hormonal Problems?

Most of us would associate hormones only with humans and more advanced animals. However, plants make plenty of these signaling chemicals. Abscisic acid helps to prepare some plants for winter dormancy. It specifically readies some plants to drop their leaves in autumn. Auxins are a group of plant hormones that have to do with cell growth in various tissues of the plant. Cytokinins are hormones that seem to stimulate cell division in most plants. Ethylene has to do with fruit ripening and shedding of leaves, among other things. Finally, a group of hormones known as gibberellins have strong influences on plant growth and flowering.

Spanish Moss

Details about this interesting plant, in the genus *Tillandsia*, are full of paradoxes. It is not from Spain and it is not a moss. Despite looking like a fungus or lichen, Spanish moss is actually a flowering plant closely related to pineapples and tropical bromeliads. Interestingly enough, it is found in parts of the Eastern United States and into South America. Spanish moss grows as an epiphyte. In other words it grows, non-parasitically, on other plants (usually trees) and absorbs water and a few minerals from the environment. Some specimens can be quite long and will hang from the branches of massive trees.

How Seeds Are Spread

In almost all cases, plants that reproduce by seeds face the problem of dispersing their seeds to a new location in order to lower competition and to spread the population. Many methods of seed dispersal have been documented and named. If plants depend mostly on water to spread their seeds, as coconut trees do, the process is called hydrochory. Wind dispersal is called amenochory. An example of a plant spreading by wind dispersal is seen in the wing like fruits of maple trees, called samaras. Sometimes a fruit containing seeds, such as a cocklebur or beggar's lice, attaches to an animal to be carried to a new location. This method is called epizoochory. Alternatively, if an animal eats a fruit and then passes the seeds in its feces, the process is called endozoochory. Many birds eat fruits and pass the seeds in this way. If a bird is the specific organism to pass the seeds, the process may be called ornithochory. A few plants disperse their seeds by mechanical means. The common ornamental touch-me-not plant, the bitter cress and the witch hazel (a small tree or shrub) propel their seeds outward as the seed pod opens.

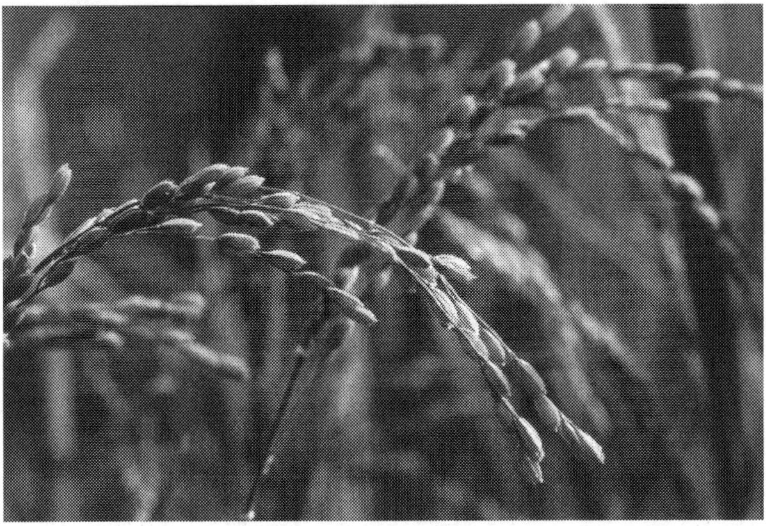

Rice, a grass, with flowering and seed bearing structures on display. Courtesy of USDA-ARS. Photographer not identified.

Grasses? Would You Believe It?

Botanists classify grasses in the family Poaceae. Many people are surprised to learn that even the grass in their yards will flower and produce seeds if left uncut for a long enough period of time. Some biologists estimate that there as many as 10,000 species of grasses. Uses for the grasses include lawn covering, food, ornamentation and even building materials. The following is list of common names of species within the family Poaceae. You may be astonished by what you find on the list: broom sedge, rye, sorghum, fescue, corn, sugar cane, river cane, wheat, bamboo, rice, oats, Bermuda grass and barley.

An Underwater Moss

Throughout much of the United States, as well as in Africa, Asia and Europe; one may often encounter moss plants of the genus *Fontinalis* growing in lakes, ponds and streams. There are several species. These dark green plants may either float freely on calm waters or be attached to rocks, sunken limbs or other substrate materials. The moss plants are often covered with sediment, giving them a brown, black or reddish coloration. They are among the largest mosses known.

A Ghost-Like Flowering Plant

Biology students all over the world learn that plants are able to make their own food because they possess chlorophyll. However, as any biology student will tell you, once you learn a rule you discover an exception. Such is the case with an unusual plant known to botanists as *Monotropa uniflora.* The common name is Indian pipe or Indian peace pipe. This plant has no chlorophyll but produces flowers. Most of the plant exists below the surface of the ground. It derives its nutrition much like a fungus. However, no fungus flowers and produces seeds. The above-ground shoots of the Indian pipe rarely exceed six inches in height. They are pollinated by a variety of insects. *M. uniflora* is often found in rich, dark woods in the southeastern United States.

Club Moss

Organisms in two genera of plants, *Lycopodium* and *Selaginella* are sometimes referred to as club moss. These usually small plants are restricted to moist conditions and are actually more closely related to ferns than to moss. They reproduce by spores, like their kin. Depending on the species and the geographic location, other common names for these plants may sometimes be heard. Examples include ground pine and running cedar. Some species, particularly those in the genus *Lycopodium,* may require nearly two decades to complete their life cycles as they change from germinated spore to spore-producing adult.

The Tree of Life

A tree native to North America, *Thuja occidentalis,* earned the nickname "the tree of life" during the 1600s. Another common name for the tree is white cedar or arbor vitae (which literally means "tree of life"). It is an evergreen, producing small cones. This slow-growing tree rarely exceeds 50 feet in height but specimens twice as tall were found in the ancient forests by early explorers. While it lives to a considerable age, sometimes up to 300 years, other trees live longer. For example, the bristlecone pine, *Pinus longaeva,* is widely regarded as being the longest lived tree on earth. Analysis of tree rings suggests an age of more than 4,800 years in some specimens. The tree of life was widely exported to Europe as an ornamental. In turn it was planted in yards, estates and cemeteries along with a related species (*Thuja orientalis)* from Asia. Although there is much speculation about the source of the nickname "tree of life," some writers have suggested that the stores of vitamin C within the tissues of the tree hold the key. It was probably used extensively to treat the deadly vitamin C deficiency called scurvy. Other medicinal uses for the tree have been recorded as well.

Whisk Ferns

The plant division Psilotophyta includes an unusual genus called *Psilotum* that provides the division with its common name, whisk ferns. *Psilotum* ferns superficially resemble a small but sparsely branched whisk broom. They lack roots but do have vascular tissues within the aerial plant parts. No leaves are found on the plants. *Psilotum* reproduces by means of spores, just like true ferns. Whisk ferns are native to some parts of the United States. Some species occur in other subtropical and tropical areas.

Specialized Leaves

In many species of plants, leaves have become specially modified in a number of ways. Genes control this altered development, and the modifications serve the plant in coping with the pressures of its environment. Some leaves or leaf stems, called petioles, are modified into structures called tendrils that help a plant to anchor and climb. Examples include leaves of pea plants, grape vines, ivy, watermelons and cucumbers. Spines are derived from leaf tissue. In some cases, leaf margins may have protective spines. The American holly is an example. Some spines are modified from stipules that grow at the base of the petioles of some leaves. In some cases, entire leaves are modified into spines. Cacti sport such leaves.

What Kind of Fruit?

A mature plant ovary, or portion of an ovary, containing one or more seeds is called a fruit. Forms of fruits range from apples and oranges to things not commonly thought of as fruits, such as nuts and berries. Botanists have given special names to these various forms. Pomes include pears, quinces and apples. Legumes are fruits of beans, peanuts and peas. A single fruit from the sunflower or buttercup is called an achene and consists of one seed surrounded by its coat. Multiple achenes make fruits like strawberries. The term caryopsis is used to describe one seed of various grains such as corn, rye, oats, barley, sorghum and rice. A tomato is a berry, as are grapes, peppers and eggplant fruits. Examples of a drupe include the peach, olive, cherry, coconut, cashew, almond and plum; multiple drupes occur as raspberries and blackberries. Watermelons, cantaloupes, squash, pumpkins and cucumbers are examples of a type of fruit known as a pepo. The term hesperidium is applied to all citrus fruits like grapefruits, lemons and oranges. Nuts are indeed fruits and include acorns, walnuts, hazelnuts and chestnuts. Other types of fruits not ordinarily thought of in terms of human consumption include siliques, which are rather like the leguminous fruits of beans. Samaras and schizocarps are fruits associated with trees like maples and elms.

Do They Really House Birds?

A beautiful and interesting fern, *Asplenium nidus*, is known as the "birds nest fern" to botanists and fern enthusiasts all over the United States. These ferns are epiphytes, meaning that they commonly grow upon trees or other suitable substrates. They make their own food and take in their own moisture. They use the trees only for resting places. The fronds, or leaves, of the bird nest fern radiate out from a central point. They may reach between two and four feet in height at maturity. The effect is rather like a large, green badminton shuttlecock. *A. nidus* may be found as a native plant in Hawaii, in parts of Africa and elsewhere. They are often grown in cultivation due to their unusual structure.

Oldest Known Living Plant

The title of "oldest living plant" is a hard title to bestow. A single specimen of the box huckleberry, *Gaylussacia brachycera,* is regarded by many botanists as the oldest known living plant and, perhaps, the oldest known living thing on the planet. This particular specimen was discovered more than 80 years ago in Pennsylvania. Most botanists who have examined it are convinced that it is a single plant that has grown more than a mile in length. *G. brachycera* is known to spread asexually by way of underground stems. At first, scientists believed they had found a population of several different organisms. However, DNA evidence reveals that the "individual" plants are genetically identical. Based on the average growth rate of the plant in other locales, the Pennsylvania specimen is estimated to be somewhere between 8000 and 13000 years old. Depending on how one views and defines an individual organism and on how age is estimated, other plant specimens (from other species) may equal or exceed this estimate.

Dahlias

Dahlias are native to Central America and to Mexico. They belong to the flowering plant family Asteraceae and the genus *Dahlia.* The plants are popular as ornamentals in many parts of the world. According to legend, they were given their common name by a European in the late 1700s to honor a botanist named Dahl. The legend states that Dahl had a hairstyle that reminded the person who named them of the plant's showy flowers. Most dahlias grown in gardens today are hybrids of two species, *D. pinnata* and *D. coccinea.*

Some Violets Don't Shrink

The family Violaceae includes nearly 900 different species of plants. In the United States, most members of the family belong to the genus *Viola* which includes those small, often showy, flowering plants known as violets and pansies. A plethora of flower colors including yellow, white, blue, purple and various multicolor combinations are known. Some people use violet flowers as cake decorations and even prepare them as a sweet side-dish known as sugared violets. Violets were once widely used by the perfume industry. It is in the tropical regions of the world that members of Violaceae attain their greatest sizes and more substantial growth forms. In fact, some species within the family grow as shrubbery, vines and even as small trees.

The Rubber Tree

Rubber was known to the mainstream scientific world as early as the 1500s. It is important to note, however, that native South Americans had made rubber for many years prior to this time. Rubber may be made by heating a type of tree sap known as latex. Latex is a milky liquid made by several trees. The best known latex producer is the rubber tree, *Hevea brasiliensis,* which grows in South America. Modern techniques have improved the strength and durability of rubber since its discovery centuries ago.

The Resurrection Plant

One or two species of the genus *Selaginella* have been tagged with the common name resurrection plant due to their amazing ability to withstand drought conditions. The genus is in the family Selaginellaceae, the spike moss family. The plants are generally restricted to moist environments and maintain a small size. During dry weather the resurrection plant turns brown and forms a dried cluster of leaves. The plant looks like it is dead. Upon exposure to adequate moisture, the leaves unfurl and the plant turns a life-sustaining green color due to its chlorophyll content. Several other species in the genus are known. They are widely distributed but most commonly are found in tropical regions of the planet. A few are used as ornamental plants.

Plants Need Their Minerals, Too

Vitamins and minerals are usually associated exclusively with animals. It turns out that plants require many types of minerals to keep their health at maximum. Plants, of course, make their own food. Yet, they usually take in needed minerals in chemical compounds that have been dissolved in water. They get important elements such as zinc, copper, iron, calcium, phosphorus, sulfur, potassium and nitrogen in this way. Most commercial fertilizer packages will list percentages of elements and/or minerals in their products.

Don't You Believe It

In many plant nurseries, garden centers and home stores you may have noticed a lush, delicate plant called "air fern." You may even own some air fern. Most people are shocked to learn that these "ferns" are not plants. They are not even alive. They do not need water or sunlight and they never grow. Air fern begins manufacture in the ocean. The skeleton-like remains of an animal, *Sertularia argentea*, are dredged up by ships, treated with a preservative and placed in a dark green dye bath. *S. argentea* is colonial organism and a member of the animal class Hydrozoa. Their relatives include jellyfish, coral and hydra. If you are looking for a nice decoration that requires absolutely no care, air fern may be a good choice. It does make a nice ornamental "plant."

Can You Really Tell The Age of a Tree By Its Rings?

Yes. The woody material making up tree trunks is dead xylem tissue. Each year a new ring of tissue is added around the circumference of an older layer. Each of these distinct layers is known as a growth ring or an annual ring. To count the number of annual rings, begin at the most central segment (pith) of the trunk or branch cross section. Continue counting until you reach the bark. The total number of rings will be the accurate age of that portion of the tree. The pith never increases in diameter. One annual ring usually consists of two sub-bands. The one of pale color is usually called springwood and the one of dark color is usually called summerwood. No wood is ordinarily added during the winter months. In addition to all of this the very oldest parts of many tree's trunks are darkened with oils, resins and other materials. These very dark regions are known as heartwood. The younger, lighter rings are called sapwood. Further, you can make some very accurate conclusions regarding the conditions in the tree's environment over the years. By comparing the relative thicknesses of the annual rings, one can make a reasonable determination about which years of the tree's history were good growth years. Abundant rain and plenty of sun usually produces thicker annual rings. Periods of drought produce thin rings. Keep in mind that some tree species routinely grow at faster rates than others and will, therefore, have thicker rings. You may also notice that many of the annual rings are not entirely symmetrical. They may be closely spaced on one side of the tree and widely spaced on others. This is a good indication that one side of the tree was exposed to more sunlight and/or growth space than the other. Consistent patterns of waves within the annual rings often indicate branch growth in that region. Damage due to lightening, fires and trauma are also very often preserved within the tree's trunk. You do not need to cut down a tree to study its rings. With an instrument known as an auger or borer you can remove an entire section of wood from one side of the tree to the opposite side, including the bark. Interestingly enough trees that grow in regions without distinct long-term winter seasons, such as those in tropical rain forests, do not display clear annual rings in cross section. Examples of such trees include ebony and jacaranda. Some of the woods have rich grain patterns, but they are not due to annual rings.

Some Facts on Flax

Flax is the common name given to members of a genus of flowering, herbaceous plants called *Linum.* Some species of the plant may grow up to 40 inches in height. Red, pink, white or blue flowers are seen among the species. Flax has been in cultivation for centuries. As early as 10,000 years ago, ancient people were known to spin thread and weave cloth from *Linum.* The durable fibers in the plant stem are still used today for these purposes. In past times, these fibers have also been used in the book binding industry. In addition to thread and cloth, many other uses for the plant have been recorded throughout history. The paper used to make money is prepared from flax. The plant's seeds yield a substance known today as linseed oil. This material has been used extensively in the manufacture of soap, paint, ink and even flooring products. The seeds themselves are still used today as a food for livestock.

Such Sensitive Ferns

Botanists call it *Onoclea sensibilis*. Early colonists called it sensitive fern. The name conjures up all sorts of images of delicate, lacy foliage or perhaps of a plant that defensively wilts when touched. These ferns are native to Eastern North America and are usually between one and one-half to two-feet tall. Some specimens may reach up to four feet, however. They are commonly found along stream banks, in swamps and in similar habitats. So, why is the common name used? It is because these plants are often among the first to die in late fall or early winter. They are sensitive to frost! Sometimes, there is no interesting story.

The Weeping Willow

The well-known *Salix babylonica*, or weeping willow, has been the subject of countless songs and pieces of poetry. It has become a symbol of sorrow and loss. In fact it was commonplace in some areas of the country to engrave an image of the majestic tree on tombstones, or to plant a tree near the grave of a loved one. It is interesting to note that this tree, which is so much a part of our history and culture, is not a native of our country. In fact, it is not even a native of the Americas. Historical records indicate that the tree was transferred from China to Europe many years ago. From Europe, it was brought to the United States. It quickly became a landscape favorite because of its unusually fast growth patterns and its graceful shape. Weeping willows have since become naturalized in many parts of the United States. The trees are very easy to propagate. Just stick a branch in damp ground and wait a few months. The United States is not without native willow species. The black willow, *S. nigra*; sandbar willow, *S. interior*; and pussy willow, *S. discolor* are examples.

Hidden Messages in Flowers

Many types of insects are probably able to sense and perceive images from the ultraviolet spectrum that humans cannot. Researchers were a bit surprised, however, when they began to view flowers under modified ultraviolet lights. They found that many showy flowers had additional colors and patterns that were not visible to humans. While it is impossible to say what things insects sense and perceive with any certainty, biologists believe that these ultraviolet patterns may serve to attract the attention of pollinating insects and guide them toward the flower's nectar and pollen stores.

Foreign Fruits

It is well known that the United States imports many, many types of fruits and vegetables from foreign countries. Some species have been in cultivation in North America for centuries. Some, in fact, have naturalized to become reproducing members of ecosystems within the United States. Peaches are widely grown in the United States. They are probably native to China. *Citrus* is a genus of plants from Asia and India. Examples of fruits from this genus include the orange, lemon, lime and citron. Pear trees came to us from Europe. Watermelons are widely cultivated in warmer parts of the United States but are native to the African continent. Another crop of note is the muskmelon, often called the cantaloupe. It is native to Africa and Asia.

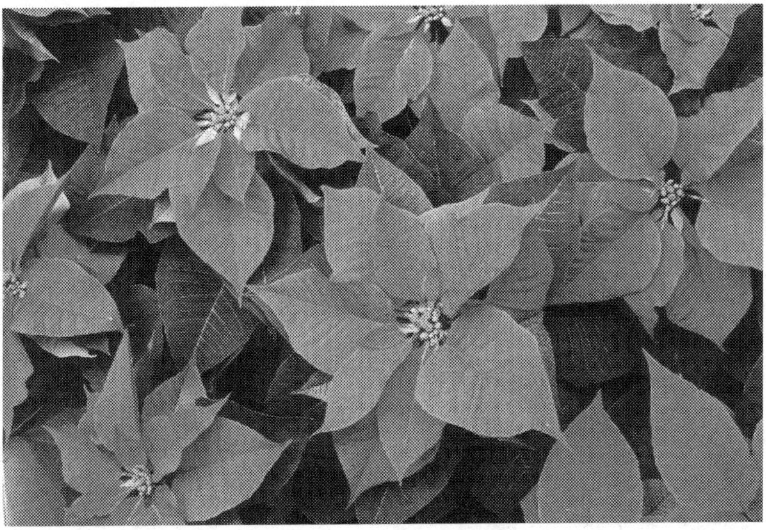

Poinsettia. Courtesy of USDA-ARS. Photograph by Scott Bauer.

The Poinsettia

Perhaps no other plant is more symbolic of winter holidays than the poinsettia. It belongs to the genus *Euphorbia*. The large colorful "blooms" most people refer to are actually modified leaves. Poinsettia flowers are actually very tiny and yellow. They are found at the center of the groups of colorful red or white leaf bracts that attract our attention. Poinsettias are very easy to maintain as houseplants but require special care to induce flowering. In order to signal flowering, the plants must be exposed to extended periods of darkness before the winter holidays. These conditions simulate the short, winter days of the plants' natural environments. According to tradition, the common name for the plant is tied to J. R. Poinsett (a prominent figure of the early 1800s) who helped popularize the plant in the United States.

Flour

This term is applied to any number of products derived from finely ground plant seeds. Most often flour is made from the seeds of various grasses including rye, buckwheat, corn and oats. The period in time during which ancient humans began to make flour remains unknown. Ancient people in various cultures used stones, wooden mallets and assorted other tools to make meal and flour from all manner of grains. In time, the power of flowing water was used to turn great stone wheels in community grain mills. Today most commercially prepared flour comes from common wheat, *Triticum vulgare*. The process begins as the wheat seeds are ground and crushed. Whole wheat flour is generally the least processed version and includes the seed kernel and husk. The production of white flour is more time consuming. It involves separating fiber-rich seed husks from the ground meal. The remaining seed material, usually of a yellowish color due to natural pigmentation, is then bleached with various chemical agents to change it to the familiar white color. Self-rising flour is a modern convenience that includes baking soda.

Why a Damp Habitat?

Have you ever wondered why mosses, ferns, liverworts and similar plants only appear in damp, shaded environments? It has mostly to do with their reproductive habits. The sperm cells of these plants are not enclosed within a protective pollen grain like those of higher plants. They must swim to the ovum within a film of water. A dry habitat would destroy all prospects for fertilization. What's more, the resulting zygote is not protected by a seed coat or fruit. It is in danger of drying out. Another thing limiting these plants to a damp habitat is the lack of vascular tissues among the mosses and liverworts. These tissues are used to conduct water and other materials. Ferns and similar plants do have primitive vascular tissues but must still rely on water for their reproductive activities.

Tulip Mania

Showy flowering plants from the genus *Tulipa* are prized by gardeners and flower-lovers throughout the world. Selective breeding has produced countless varieties of color. These flowers are native to Asia and some regions of the Mediterranean, not Holland as commonly believed. People of Holland, however, were among the first to show an interest in growing and cultivating tulips. Historians note that their interest approached mania. In the 1600s, some people actually drove themselves into bankruptcy by purchasing particularly rare and unusual tulip bulbs.

Resurrection Fern

"Resurrection fern" is a common name applied to *Polypodium polypodioides.* This fern is unusual for several reasons. It is found in Africa and in South America. In the United States, it is particularly common in the Southeast. Depending on the weather, resurrection ferns may look like a crumbling, brown mass of dried plant material. When conditions are moist, however, the ferns turn to a lush green color. Resurrection fern is an epiphyte, meaning it commonly grows upon tree trunks but still makes its own food. It is not a parasite. Resurrection ferns have long, creeping rhizomes that give them a vine-like growth habit. They most commonly occur on hardwood tree trunks.

Larch

A genus of conifers, *Larix*, is noted for strong wood and for the fact that the majority of species are not evergreen. Unlike most other conifers, the larches drop their needles during cold weather. One species in North America is *L. laricina.* It tends to grow near swamps or other damp environments. In the United States, it is restricted to the northern states. As winter approaches, its needles change to a yellow color. It is also known by the common name tamarack, and produces some of the smallest cones of any tree in North America. Another species, the Western larch, *L. occidentalis*; also grows in northern parts of the United States. Two other deciduous conifers are of note. They are the dawn redwood, *Metasequoia glyptostroboides*, and the bald cypress, *Taxodium distichum.*

The Mightiest Moss

Mosses are small plants. Their size is limited by a number of factors, not the least of which is the absence of highly efficient and specialized water conducting tissues found in trees and so many other plants. The word giant can certainly be relative. There are a few species of giant mosses known throughout the world. Some routinely exceed one foot in height. The trophy for the tallest moss in the world would be awarded to *Dawsonia superba*. This species is known mostly from New Zealand but populations are found in a couple of other places. Under ideal growing conditions this monster can sometimes reach just less than 20 inches in height. At first glance a stand of large *D. superba* may be easily mistaken for pine tree seedlings.

Bamboo

Many people are surprised to learn that bamboo is a grass, belonging to the plant family known as Poaceae. In the past, bamboo was mostly restricted to tropical and subtropical regions of the globe. Today it is in cultivation elsewhere. Several hundred species are known. They range in size from less than one foot, to more than 100 feet, in height. The plant stems are mostly hollow. Some species flower and produce seeds only once every few years. A few species, in fact, are believed to flower no more often than once every 100 years. One species may grow up to three feet in height within a 24 hour period. *Dendrocalamus brandisii* is probably the tallest species. Individual plants have been measured at around 100 feet in height and eight feet in diameter. Another very interesting species is *Phyllostachys nigra*. It is sometimes called black bamboo because its stems darken with age and light exposure. *Bambusa arundinaceae* is probably the most common species on the planet. Throughout time, bamboo has been put to a wide variety of uses. A few examples of these many uses include drinking vessels, smoking pipes, nets, construction scaffolding, caskets, medicinal potions, boats, fishing poles, bridge building, hats, paper, sleeping mats, home construction, plumbing, weapons and baskets.

About Coffee

Coffee is a pleasure drink enjoyed daily by millions upon millions of people throughout the world. Most people probably never stop to think about how their drink came to be. More than 30 species of plants in the genus *Coffea* have been described in the warmer parts of the globe. Only about three tree-sized species are especially prized and cultivated by the coffee industry. Many historians believe that these trees have been in active cultivation since the 600s. Most coffee trees reach a mature height of 12 to 20 feet. After flowering, coffee trees produce a green fruit that slowly changes color from light red to dark red or purple as it ripens. The seeds within these fruits are the ultimate source of coffee. Many people refer to the seeds as beans. They have a green color when harvested. When the seeds are roasted, they change to dark brown or black in color. When added to water, the ground and pulverized seeds yield the flavored liquid we call coffee. As a drink coffee has little to no real nutritional value. It is enjoyed hot by most people. Others prefer iced coffee.

Tallest Tree

The heights of trees are often exaggerated and, unfortunately, poorly documented. *Eucalyptus regnans* is a species from Australia going by common names such as "Victorian ash" or "mountain ash." There are numerous reports of these trees being cut, or found down, with lengths of more than 400 feet. One estimate was 500 feet. Many regard these records as credible, others do not. One of the best documented tallest trees was a Douglas fir tree that was cut in British Columbia around 1940. It is said to have measured about 415 feet in length. As a species, redwoods (*Sequoia sempervirens*) are among the trees that routinely grow to great heights. One individual was measured at 385 feet tall and had a trunk that was nearly 30 feet in diameter. It is said that a cross section of a redwood tree trunk was used to make a table that could comfortably seat 40 people at mealtimes.

What Is A Hornwort?

Hornworts are a group of small, non-flowering plants that are close relatives of mosses. They belong to the class Anthocerotae. Chances are that you have probably seen a hornwort but not really looked at it carefully. The structure of hornworts is, superficially, unremarkable. They are usually just a slender, green, stem-like structure growing from a button-like base. Hornworts almost never reach more than three or four inches in height. Most are only a fraction of that size. Among many other reasons, a noteworthy feature of their cellular structure makes them of interest to biologists. Hornworts have unusually large chloroplasts when compared to other bryophytes. Chloroplasts are specialized cell parts that help plants (and some other organisms) to capture energy from the sunlight to make food. Up to nine genera of hornworts are recognized, depending on the taxonomic treatment. Examples include *Megaceros* and *Phaeoceros.*

Chestnut of the Hundred Horses

Probably the largest tree, in terms of its trunk circumference, is a European chestnut found in Sicily in the late 1700s. It was nicknamed "the chestnut of the hundred horses" because, according to tradition, one hundred men on horseback were able to gather beneath its massive and spreading crown. The trunk measured almost 200 feet in circumference. Many scientists believe the tree is actually composed of a few smaller trees that fused together during growth. Others think it to be a single tree.

What is a Pine Cone?

Pine trees belong to a group of plants known as the Division Coniferophyta that do not produce flowers. These plants produce cones that bear pollen and ova, usually on separate structures. The woody structure we know as a pine cone is actually a female cone that has been fertilized by sperm-bearing pollen, carried by the wind. Pollen grains are released from much smaller, non-woody, male cones. Within the woody female cones are seeds, borne upon scales, which will scatter in the wind when mature. Some pine trees require two years for the seeds to mature to the point of release. In certain species the cones may persist upon trees for decades.

Viruses Make Plants Sick

We tend to associate viruses exclusively with human illness. Humans suffer from such viral induced illnesses as the flu and common cold, among others. However, plants are also susceptible to viral attack. A number of plant viruses are known. Plants may become infected by way of contact with contaminated soil and by contact with parasitic plants. A number of animals, especially insects, may also spread viruses from one plant host to another. Mosaic viruses are a group that may infect several species of plants. They attack plant cells to reduce the amount of chlorophyll in the leaves. This causes the characteristic mottling of plant leaves seen in a mosaic virus infection. An example is the tobacco mosaic virus which can damage leaf tissues and kill tobacco plants. Plants routinely suffer from tumors, some of which are likely caused from viral infections. Candy striped tulips, with their showy red and white flowers, are often planted in our yards. The flowering pattern is the result of a viral infection. Another group is the yellow viruses. They attack the vascular system of plants that is responsible for conducting water and food throughout the plant body. These viruses may cause stunted growth and curling of stems and leaves. Tomatoes are vulnerable to specific kinds of viral infections known as curly top and spotted wilt. Specific examples of other plants that are vulnerable to viral infections include carnations, cucumbers, lettuce and corn.

They're Not Ferns At All

True ferns do not produce flowers, fruits or seeds. They are limited to damp habitats because they require water in order for reproduction to succeed. Several plants have been incorrectly called ferns throughout history. *Asparagus densiflorus*, commonly called the asparagus fern, is one example. This plant is actually native to Africa. It produces tiny flowers that often go unnoticed among the lush, green foliage. Another member of the genus, *A. setaceus* is often referred to as the lace fern. Sometimes the common names are exchanged between these two species. An additional flowering plant with the term fern in the common name is sweet fern, *Comptonia peregrine*. This plant is native to the United States.

Compound Flowers

Some plants produce compound flowers for reproduction. It is surprising to learn that some things we refer to as a flower are actually collections of multiple flowers, sometimes called composites. The sunflower is an example. Large plants may have hundreds of tiny flowers grouped together on a single stem. Those near the edge of the cluster may have elongated petals and are called "ray flowers." Those within the circle of ray flowers lack highly elongated petals and are called disk flowers. Other examples of composite producing plants include daisies and asters.

A Non-Bostonian, Actually

Several ornamental ferns have been given the common name Boston fern. None of them are native to North America. The fern known to botanists as *Nephrolepis cordifolia*, is one of those most frequently called Boston fern. Reasons behind this widely used common name are unclear. As a group, these plants were among some of the very first ferns to be widely grown in cultivation in the 1800s. Even today, Boston ferns are often found in hanging baskets in and around many homes in the United States. In at least two or three states, especially in Florida, these organisms have escaped from cultivation and have become naturalized. They are considered to be pest plants in these areas. If these escapees were put in hanging planters, trimmed and placed in a greenhouse, they may fetch a nice price.

Variegated Plants

One may find variegated plants in the wild on occasion. Variegation refers to the presence of a striping or mottling pattern on the plant leaves. Plant nurseries sell many varieties that have been propagated from a naturally occurring plant or that have been encouraged through selective breeding. The key to variegation in most plant species lies in some specialized cell parts known as plastids. Plastids are organelles that generally have something to do with food production or food storage. There are several types, each containing characteristic pigments. The most well known plastids are chloroplasts which contain the green pigment chlorophyll. On occasion, plastids may be defective due to a mutation. These plastids may lack their characteristic pigment. During cell division, some of these defective plastids may cluster in large numbers into one daughter cell. Subsequent cell division may produce segments of leaf tissue wholly or partially devoid of the traditional green pigmentation.

A Rose by Any Other Name

The roses, familiar to most everyone, belong to the family Rosaceae. This extremely diverse family, recognized by common features of fruit and flower structure, is not limited to just roses. More than 2500 species of plants (including some trees) are classified alongside roses within the family. Examples of members of the rose family include blackberries, *Rubus*; cherry, *Prunus*; apple, *Malus*; strawberry, *Fragaria*; and many types of herbs and wildflowers.

Do Plants Breathe?

Absolutely! Most land plants have tiny microscopic openings in their stems and leaves that are called stomata. The singular form of the word is stoma. These pores are usually opened and closed by the action of two guard cells which form a gate-like covering across the pore when they are swollen with water. Respiratory gases enter and exit the openings. Many people believe that plants only take in carbon dioxide and release oxygen as a byproduct. This is the case when the plant makes its food by photosynthesis. However when plants turn their food into energy, they take in oxygen gas and release carbon dioxide just like animals. Water vapor is also released through the plant's stomata.

Money Making Moss

Mosses in the genus *Sphagnum* constitute one of the most economically important groups of plants on the planet. There are several species that occur around the globe, mostly in the Northern Hemisphere. Especially high concentrations of the plant are known in Siberia, Florida, Wisconsin, Canada, and in other areas. The plants go by common names such as sphagnum moss, peat moss and bog moss. In some places they grow in small, isolated stands. In other locales they form great dense masses, called hummocks, in bogs or swamp-like environments. In growth situations such as these the moss presents with lush, green growth above the surface of the water. Below, it ranges from brown to black in color reflecting various states of decomposition. Such large concentrations of sphagnum are often referred to as peat bogs. One of the most striking things about sphagnum or peat moss is its ability to hold water. Even dead peat moss can soak up and retain as much as twenty times its own weight. This is possibly due to the presence of specialized barrel cells within the tissues of the plant. Throughout history, sphagnum moss has been used extensively by humans in great numbers of ways. It has some antibacterial properties, and has been used to bandage and dress wounds. Dried peat moss makes a good fuel. The plant has been used to chink cabins and continues to find a great deal of use among florists and gardeners. It is also of note that peat bogs provide habitats for some species of carnivorous plants.

Edible Flowers

Several flowers are edible. Examples of flowers that you may regularly eat without realizing that you are consuming flowers include broccoli, cauliflower and artichokes. Some people may eat the showy flowers of other plants or use them for accenting and decorating other dishes. Flowers such as these have been slower to find their way into mainstream dietary habits.

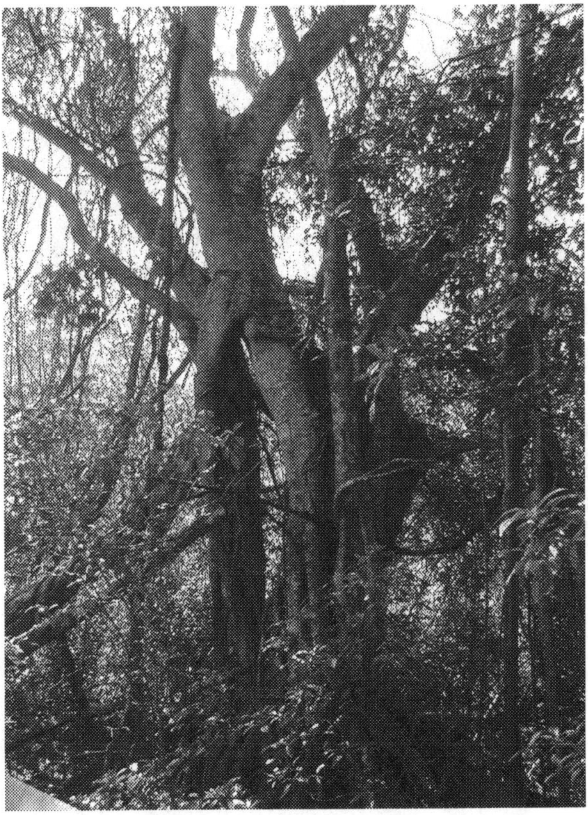

Strangler fig on host tree. Courtesy of USGS.
Photograph by Heather Henkel.

One Tree Murders Another

An unusual tree called the strangler fig (*Ficus aurea*) is native to a few tropical and semi-tropical regions. The tree is tall and slender, almost superficially vine-like in its growth patterns. It begins to grow high in the crown of its host tree from a small seed. Seeds are usually transported with bird feces. The developing fig trees send out adventitious roots. Once they make contact with the ground they thicken and encircle the host tree. In time the host tree usually dies and rots away, leaving the fig's hollow trunk-like root system.

Cocoa, Chocolate and Cocoa Butter

Many are surprised to learn that chocolate and related products come from a tree that is native to South America. *Theobroma cacao* is the scientific name for the cocoa tree. It has been utilized as a food source for centuries by native people and is widely cultivated in other parts of the world. The story begins with the fruit of the plant. Seeds within the fruit are fermented by bacteria and yeast. The resulting product is dried and ground into what many refer to as "bitter chocolate." If sugar is mixed with the bitter chocolate, sweet chocolate is the result. Milk chocolate is derived from the addition of sugar and milk to the bitter. Depending on the desired result, the chocolate may be processed further. Fat that is removed and purified at this stage is known as cocoa butter. The remainder may be dried and ground into a powder known as cocoa.

Specialized Plant Stems

Botanists classify certain specialized parts of plants as stems. Superficially, some examples are not very much like traditional plant stems. For example, plants such as the iris, begonia, and bamboo grass have fleshy, horizontal, underground stems that are known as rhizomes. The same sort of organ may occur above ground in plants such as the strawberry and spider plant. They are known as stolons or runners and help the plant to reproduce asexually. Stems on roses, hawthorns and other plants are sometimes modified into thorns or prickles. Thorns are often woody; prickles do not involve woody tissues but are made from the plant's epidermis and sub epidermis. Thorns are rather tough and prickles are easily broken away from the stem. Rose "thorns" and those of blackberries and raspberries, are actually prickles. A bulb is actually a stem bud surrounded by fleshy leaves. Examples may be found in the tulip, lily, daffodil and onion. Tendrils, such as those found on grape vines and other plants, help the plant stem to increase in length toward the sun by providing anchorage and support. Some tendrils are derived from leaf tissue.

Liverworts

The common name liverwort is given to a large group of plants within the division Bryophyta. Like their relatives, the mosses, liverworts tend to be relatively small. They lack vascular tissues and seeds, and are restricted to moist environments. Liverwort plant bodies (also called thalli) are often incised into deep lobes. This reminded ancient people of the human liver. They assumed that the plants were valuable in treating liver ailments. The name has stuck through time. Liverworts may reproduce asexually by fragmentation. Pieces of the body may be broken away and will often begin growing on their own. Some species produce unusual asexual reproductive structures known as gemmae. These are tiny clones of the parent plant that are held within a cup-shaped cavity on the parent's body. Raindrops serve to scatter the gemmae away from the parent when they strike the cavity. Sexual reproduction is achieved by male and female gametophytes that produce sperm and ova. After fertilization, sporophyte plants undergo a specialized type of cell division known as meiosis so that new gametophytes may be produced. Some species' spore capsules have spring-like tissues called elaters. They twist and shift dramatically to help the disperse spores over a wide area.

Plants with Veins?

Plants do not have blood or a heart but some of them do have a system of vessels that are collectively called veins. Vascular plants (those that possess these sorts of tissues) include trees, many wildflowers, ferns and related species. Two basic types of tissues form these vein-like canals. Phloem forms vessels that transport food, or products made from food, throughout the plant. Xylem forms vessels that transport water and dissolved minerals. An entire group of nonvascular plants exist that lack the tissues described above. They rely on processes such as diffusion, osmosis and streaming of the cellular cytoplasm to circulate food and water. Examples of nonvascular plants include the liverworts, mosses and hornworts.

Wheat: A Plant with Many Uses

Wheat is a grass of the genus *Triticum* that has been in cultivation for centuries. Wheat products have been recovered from tombs in ancient Egypt. Evidence for even earlier use of wheat by humans is also known. The species known as *T. vulgare*, or common wheat, was probably brought to colonial America as early as the 1500s. This species is hardy and rarely reaches more than four feet in height. Wheat has been used as animal feed, for making mats and baskets, and as food for human consumption in various parts of the world. Flour derived from the pulverized seeds may be used in making bread, pasta, a coffee substitute and other foods.

That Reminds Me of Cinnamon

One of the largest ferns in the continental United States is the cinnamon fern, *Osmunda cinnamomea*. Given ideal growing conditions, older specimens of the organism can top six feet in height and have a width of three to four feet. The fern is commonly found near springs, bogs and other damp locales in many parts of the United States. The common name is derived from the cinnamon-colored fibers near the plant base, and from the similarly colored spores of the fern. Above-ground portions of the plant die back each year. The rhizome (underground stem) and roots may survive for decades.

This Fern Gets Around

The United States is home to an unusual and rarely noticed fern that has been given the common name "walking fern." The scientific name of the organism is *Camptosorus rhizophyllus*. Some botanists place it in the genus *Asplenium*. The arching fronds or leaves of this fern are lance-shaped, with a pointed tip. They sometimes reach one foot in length, but are usually much smaller. If the tip of the fern is in contact with damp ground long enough, it will produce roots to give rise to a new plant, genetically identical to its parent. This is how the name walking fern originated. Over time, the plant appears to be moving about. The organism can reproduce sexually as well.

Jojoba Oil

Jojoba oil is a relatively recent commercial product. It is derived from the seeds of a shrubby plant, known to botanists as *Simmondsia chinensis*, which grows in Mexico and in the Southwestern United States. The hardy plant can tolerate desert conditions. Jojoba oil has been known to native people for centuries. However, it is only within the last several decades that the oil began to be commercially marketed on a large scale. The oil is valued as a lubricant and is often added to cosmetic products such as soaps, shampoos and lotions.

Plant Pigments

Most of us associate the green pigment known as chlorophyll with plants. In addition to providing coloration, chlorophyll helps plants to capture energy from sunlight to make food. However, many other pigments may be extracted from green plants. These include yellow to orange carotene pigments, red or purple anthocyanin pigments, red and blue phycobilins, and yellow xanthophyll pigments. Many of these pigments are masked by chlorophyll in green plant leaves during the spring and summer. They become visible in fall, accounting for the array of colors commonly observed in autumn leaves. It is of note that some pigments may appear in spring and fall in other plant parts. For example, carotene pigments provide color to tomatoes, squash and banana fruits, as well as the roots of carrots.

Use a Horsetail to Scour the Pots

One of the most curious of the primitive plants is known by the scientific name *Equisetum hymenale.* These evergreen organisms are composed of hollow, jointed stems that very rarely branch. The stems may reach up to sixty inches tall but are usually not much taller than a foot. They are usually topped by a spore-producing structure. The stems have high silica content. In olden days handfuls of the rough stems were used to clean dishes, polish metal and smooth wood. The plant has even been used to make paper on a small scale. This diversity of uses for *E. hymenale* has earned it a string of descriptive common names that include horsetail, scouring rush, bottle brush and pewterwort.

Those Ancient Cycads

Only about 100 species of plants in the family Cycadaceae are known to biologists today. Several fossilized remains of plants from the Triassic period of the Mesozoic era are classified as extinct members of the family. Some botanists have called this time period (about 250 million years ago) during which the group flourished, "the age of cycads." Today, cycads grow as shrubs or small tree-like organisms and are most abundant in South America. The tallest plants rarely exceed 50 feet in height. They are classified as gymnosperms (along with pines, cypresses and other plants) due to the fact that their ova or egg cells are not covered by a coating as in flowering plants. Some species bear female cones within which seeds mature after fertilization. Cycads have leaves that resemble those of palm trees. Their male reproductive cells, called sperm, possess numerous whip-like flagella for locomotion. Some genera such as *Zamia* may be grown as ornamentals. A specimen of the cycad *Encephalartos altensteinil* was collected in South Africa in 1775 and placed in a pot. As of this writing, the plant still survives in Great Britain.

What Long Leaves You Have

As a group, the raffia palms must have the longest leaves of any tree. These palm trees, which are not particularly tall, are found in Africa and Madagascar. They are placed in the genus *Raphia* and may regularly produce leaves in excess of fifty feet in length. The champion member of the species must be *R. regalis*. A few reports of leaves that exceeded 70 feet have been filed. The most extreme claim, 82.5 feet, is not unlikely. Another noteworthy thing about *Raphia* is raffia fiber. It is harvested from the trees in great quantities and sold commercially. It has been used to make cloth, baskets, hats, twine and other useful items.

Staghorns

Several species of ferns, mostly in the genus *Platycerium*, are often referred to as "staghorn ferns." Some fern growers would split hairs on details, but the common name "elkhorn fern" may also be used for members of this group. The large fronds of these plants may remind one of green antlers. The ferns are mostly native to tropical regions of Australia, Singapore, South America and Madagascar. In such habitats, these ferns are often found growing from the crotches of trees. They use the tree as a host but do not take any water or nutrition from it. Instead, the tree provides a space for anchorage and growth. *P. bifurcatum* is one species that is commonly grown as a houseplant. Specimens may be mounted on wooden or metal supports. As the plant grows, the green fronds will hang down in the same way as they would in their native tropical environment.

Water Lilies

These beautiful plants, romanticized in song and poetry, belong to the family Nympaeaceae. They are also called as pond lilies. More than 50 species are known, although some exist only in cultivation. A few types are found in the United States. The water lilies attain their greatest diversity, however, in tropical regions of the globe. The showy flowers ordinarily rise just above the surface of ponds and slow-moving streams. Flower color depends on the species. Many shades of white, pink, yellow and other colors are known. Most of the leaves are beneath the water until flowering occurs. The plant is anchored by way of an underground stem. At one time, many wealthy people grew water lilies as a hobby. Today the plants are often featured in man-made fish ponds. One species from South America produces stout leaves that sometimes measure more than six feet across. Some species have floating leaves strong enough to support the weight of a child or small adult.

Glowing Moss

A few genera of luminescent mosses have been described by botanists. One of the best known species is *Schistostega pennata*, commonly called "cave moss." A few other unusual common names for the species are known. They include goblin's gold, elf gold and luminous moss. The green-yellow glow of *S. pennata* is emitted when light strikes the tiny plant's leaves and is reflected back from the surface. Since they are commonly found in caves and other very dark places, the effect can be eerie. Stories and legends about *Schistostega* are common in Japanese culture. The group is also known from Europe, and parts of the United States and Canada. The organisms are less than one-half inch tall.

What about a Cactus?

The common name cactus (the plural is cacti) is widely applied to untold numbers of plants that grow around the world. In a more formal sense, members of the botanical family Cactaceae share several things in common. In most species, the leaves are much reduced to form spines. Most of the green chlorophyll the plant uses to make food is found in the plant stems and buds. Cacti produce flowers, often very colorful and showy, for reproduction. Seeds are usually found within berries or burs. Some of these fruits are valued as food by humans and other animals. The size of mature cacti varies from only fractions of an inch in height to tree-size forms (like those so often associated with the old West) that can surpass 60 feet in height. Cacti are not limited to desert-like environments. Some species occur in the temperate regions of the Eastern United States and others are found in rain forests.

Plants Respond to Their Environment

Plants display a variety of growth responses, called tropic responses or tropisms, to stimuli in their environment. If the plant grows toward the stimulus, it is said to be a positive tropic response. If it grows away from the stimulus, it is called a negative tropic response. Different parts of the same plant may display different responses. For example, roots tend to avoid light (a negative phototropism) while plant stems and leaves tend to grow toward light (a positive phototropism). Likewise, roots display a positive gravitropic response while stems and leaves generally display a negative one. A gravitropic response, by the way, is a response to the pull of gravity. The term skototropic response emphasizes a plant's response to darkness. We would expect plant roots to be positively skototropic and leaves to be negatively skototropic.

Vanilla

Numerous vanilla products, including natural vanilla flavoring, are derived from a vine-like flowering plant. *Vanilla planifolia*, an orchid mostly found in Mexico and South America, produces dark flavor-laden seed capsules from which vanilla is extracted. The seed pods are often wrongly referred to as "vanilla beans." Artificial vanilla flavoring is made in chemical laboratories. The flavor closely matches that of naturally derived vanilla.

The Filmy Ferns

The plant family Hymenophyllaceae includes several species that are tagged with the common name "filmy fern." These extremely delicate organisms are found mainly in the Southern hemisphere. Many species grow upon the trunks of trees in rainforest environments. Their common name derives from the fact that many of their leaves or fronds are only one cell layer thick. The most common genus is probably *Hymenophyllum.* Another genus, *Trichomanes*, has species that occur in New Zealand, Australia and in the United States. *T. boschianum* is found in only a few scattered locations in the Appalachian Mountains and is commonly called the "Appalachian filmy fern." In some locales it produces lush, but fragile, lacy fronds. In other populations, it exists only in its gamete-producing stage that looks very un-fern like to the average person. A second species in the genus, *T. intricatum*, grows almost exclusively as a mass of green thread-like structures.

CHAPTER THREE

Kingdom Animalia: The Invertebrates

What Do Spiders Eat?

Spiders are carnivores, meaning that they eat the tissues of other animals. All spiders, with the exception of the family Uloboridae, have glands which producc venom. Spiders use this poison to kill prey and defend themselves. Most spiders spin capture webs to catch insects or similar food items. The webs of some species may include sticky glue-like beads which help entangle the prey item in the web. *Frontinella pyramitela*, a spider common in the United States, has often been called the "cup and doily spider" because its capture web consists of a relatively flat area of silk upon which a cup-shaped region rests. Some spiders do not spin capture webs but actively stalk, or lie in wait for, their prey. Spiders called spitting spiders, in the family Scytodidae, eject a mixture of venom and sticky mucous to capture their prey. In addition to insects, some species of spiders may feed on small fish, birds, other spiders, small frogs and a variety of other types of food. One family of spiders, Mimetidae, contains species that apparently live within, and steal food from, the webs of other spiders. This habit has earned them the nickname "pirate spider." In all cases, spiders utilize digestive juices to break down their prey.

Leech Facts

Not all leeches are blood-sucking parasites. Some leeches feed on algae. Others are carnivores and may eat small worms and similar prey. Some species feed only on dead and decaying animal matter. Almost 700 species of the organisms are known. While most leeches live in fresh water, some reside within especially moist soil. Leeches belong to the same taxonomic phylum (Annelida) as earthworms but are in the class Hirudinea. Those leeches that do feed on blood secrete a chemical call hirudin that slows or prevents blood coagulation in their host. Additional chemicals may numb the feeding site. In this way, the leech can continue feeding from a wound made with its sucker.

Scanning electron micrograph of a mite.
This species is a carrier of a virus infecting citrus trees.
Courtesy of USDA-ARS. Photograph by Eric Erbe.

A Mite Might Trouble You

Mites are tiny relatives of spiders and are classified with them in the class Arachnida. Hundreds of species have been described. The feeding habits and life histories of mites are extremely varied according to their species. Some exist within leaf litter or soil. A few species inhabit the dung piles of other animals. Many species of mites are parasitic. They are known to parasitize a number of plant and animal hosts. Those mites make their living by feeding on plant sap, animal secretions or blood. Some species are tiny enough to seek out insect hosts. Bees, caddis flies and many other kinds of insects unwillingly pick up the checks of mites. They also infest birds, snakes, humans and other species of mammals.

Hydra

Hydras are among the simplest animals known. They are tiny creatures that are commonly found in fresh water ponds and quiet streams, clinging to rocks or leaf litter. Hydras belong to a phylum of animals known as Colenterata. The phylum also includes coral, jelly fish and the Portuguese man of war. Just like its cousins, the tiny hydra has tentacles equipped with stinging cells called nematocysts that it uses to capture and stun prey. The prey is carried to the mouth for digestion. Although hydras spend most of their time attached to a substrate, with their tentacles positioned like points of a crown, they can move. Hydras can turn flips, end over end, to travel to a new location. Hydras may also move by gliding along the surface of a substrate.

Paper Wasps

Many people in the United States are familiar with paper wasps. These members of the genus *Polistes* build nests of chewed plant-based materials upon which their colonies are based. The nests are frequently found in and around human houses and tend to be gray in color. Nests usually hang from a single stalk and vary in size. In the early spring a small group of sisters emerges from hibernation. They begin to build a new nest for the season. All of the *Polistes* sisters mated with a male from another colony before entering hibernation the previous autumn. These female paper wasps begin to lay fertilized eggs within the cells of the growing nest. One of the sisters becomes the colony queen and begins to devour the eggs and larvae of her sisters. The sisters help to protect the nest. They become sterile as their ovaries degenerate. As the colony grows, all other offspring are produced by the queen. By late summer and early fall, male offspring begin to be produced from unfertilized eggs. Mating takes place between the members of various colonies. The females that mate enter hibernation in late fall. They will begin the new colonies early the next spring.

Angling Spiders

An entire family of spiders, Pisauridae, has been tagged with the common name "fisher spider." This is because some of the larger species are often observed feeding near streams, lakes and ponds. They routinely capture water insects and even small fish. The genus *Dolomedes* is regularly found in such habitats within the Eastern United States. Even though fisher spiders frequent damp habitats, they are strictly land creatures. They are unable to swim and can remain submerged in water for only very short periods of time. If they do come into full body contact with water, it is usually by accident. As a general rule the fishing spiders, including *Dolomedes,* try to keep a foot hold on the shore or on floating debris while capturing their prey. Because of their affinity for water, *Dolomedes* often find their way into people's homes by way of shower drains and the like. A mature *Dolomedes* is a large and impressive spider specimen. Some people mistake them for tarantulas. Their venom is not generally toxic or life threatening to humans but when the spiders sense the need to defend themselves, they can deliver painful bites.

Deadly Lady Bugs

Those tiny, cute, ladybugs so often seen during warm weather are familiar to everyone. What the average person may not know, however, is how very important these insects are to the ecosystem. Like all other beetles, these bugs (also known as ladybird beetles) belong to the order Coleoptera. Several species are known. The larva of one genus, in particular, is known for its ability to kill and eat aphids, tiny organisms that parasitize plants. Members of the ladybug genus *Hippodamia* have larval stages that may reach about one fourth of an inch in length. These hungry larvae can consume more than three dozen aphids within an hour's time.

Fish with Shells?

The term shellfish has been commonly applied to a very dissimilar group of aquatic animals. The name, however, is problematic and misleading. It has no actual basis in the biological classification of these organisms. In fact, the group known as shellfish includes organisms from more than one phylum. Oysters, clams, scallops, crabs, lobsters and even starfish are sometimes grouped together into this artificial cluster. The term shellfish appears to be more associated with the food industry than with biology.

Millipedes and Centipedes

Many people know these organisms by their respective common names, "thousand legged worm" and "hundred legged worm." However, as is often the case, common names can be deceiving. In the first place neither of these organisms are worms at all. They belong to the phylum Arthropoda which includes the insects, spiders, lobsters, crabs, mites and many other animals. The millipedes belong to the class Diplopoda and usually have two pairs of legs per external body segment. Even the largest and most complex species of millipedes have fewer than 100 legs. Most species of millipedes are harmless plant eaters. They often begin their life with only three pairs of legs. The number of legs may increase with molting as the organism grows. Some release a toxic odor to discourage predators. The centipedes belong to the class Chilopoda and have just one pair of legs per external body segment. In the largest and most complex species, some centipedes may in fact have more than 100 legs. Most species have fewer than 100 however. Some centipede species are active hunters and secrete a toxin from structures near their mouths.

Silk Comes From Where?

Many different kinds of organisms, including spiders, produce silk. The silk cloth that most humans know is derived from the silk moth industry. As far as historians can surmise, the culture of silkworms and the silk industry probably originated centuries ago in China. The oldest known species of silkworm cultivated for industrial purposes is *Bombyx mori*. These organisms are entirely unknown in the wild today but are extensively kept in culture around the world. The silkworms, which are actually moth larvae, feed mostly on mulberry leaves until it is time for metamorphosis to begin. When the time is right, they spin a single thin silken thread to form a cocoon. If left undisturbed, an adult silk moth will emerge from the cocoon within several days. Before this stage, however, silkworm farmers will kill the worm inside the cocoon. They carefully dismantle the cocoon by spooling the long silk thread around a stick or similar structure. The threads may then be woven into fine cloth.

Sextons Officiating at a Burial

Sexton beetles are small, dark colored insects that rarely reach lengths of one and one-half inches. All known species belong to the genus *Nicrophorus*. These creatures are noted for their remarkable ability to bury comparatively large animals like rodents, birds and snakes for purposes of feeding and reproduction. Sexton beetles work in pairs, male and female. They locate a small dead animal and excavate beneath and around the carcass, effectively lowering it into the soil. Once the beetle parents begin their work, they remove most all traces of feathers or hair from the carcass. When the animal is completely buried, the beetles mate. The female deposits her eggs near the dead animal, beneath the soil. The parents will also construct a passageway for the escape of their offspring from the grave. Once the fertilized eggs hatch, the larvae begin feeding on the remains of the dead animal their parents provided. They next enclose themselves in a pupal case and then emerge as adult beetles. The cycle begins again.

Army Ants

Eciton burchelli is the scientific name of the army ant. These organisms have long been talked about and written about by explorers and biologists alike. The species is from South America. These ants live in colonies that range from one half a million to about two million in size. As in so many other ant colonies, a single queen produces larvae and the workers cooperate to find food. A large army ant colony is so efficient at food getting that it acts almost like a single large predatory organism. Early in the day, long streams of workers exit the colony in raiding columns. These columns expand laterally to produce great swarm fronts. As the entire colony moves, workers cooperate to capture all sorts of prey items that are shared in the colony but mostly used to feed growing larvae. As night draws near, the total colony assembles into a ball of workers that surrounds and protects the queen and larvae. This massive sphere of ants is often called a bivouac and may rest among tree branches or large rocks. Very few small organisms can survive the onslaught of an army ant colony. Spiders, insects and other small animals are no match. Some extraordinary stories about large mammals succumbing to a hungry army ant colony have been widely circulated. Most biologists view stories involving humans, livestock and other large organisms with skepticism.

Fish Eating Beetles

More than 3000 species in the insect order Coleoptera are known as diving beetles. Some grow to more than one and one half inches long. One species native to Europe and Asia may approach two and one half inches in length. The adults and larvae of diving beetles are often found in ponds and slow streams. They are active hunters and feed on a variety of organisms including other insects and even small fish. In some species the larvae go by common names such as water tiger.

Spinning Spiders

All known species of spiders make silk. Some species, in fact, may make several different kinds. Spider silk is mostly protein based. The protein is similar to the type that makes human hair and fingernails. Spider silk begins as a liquid. It is produced by a gland within the spider's body and released through an opening known as a spigot. Biologists still do not fully understand how the spider is able to transform this liquid silk into a thread or fiber. They do know that spiders have two, three or four pairs of appendages at the rear of the abdomen that are known as spinerettes. These organs are somehow involved in manipulating the silk threads. Spider silk is extraordinarily strong and most types are very elastic. Use of silk varies highly by spider species. Spiders may produce silk for purposes of making capture webs, lining burrows, protecting eggs, and for any number of other reasons.

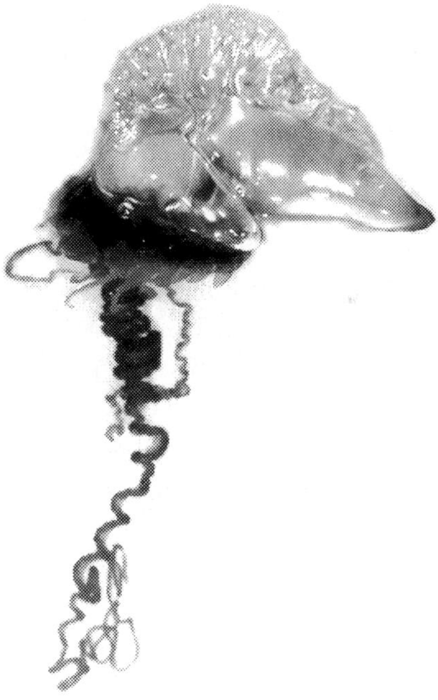

Small Portuguese man-of-war colony. Courtesy of National Oceanic and Atmospheric Administration/Department of Commerce (NOAA). Photograph by Bruce Moravchik.

Portuguese Man-of-War

Although they belong to the same phylum (Coelenterata) as the jellyfish, the Portuguese man-of-war (sometimes called man-o-war) is not a true jellyfish at all. There are some other very interesting items of note about these organisms. First, the man-of-war is not a single organism but actually a highly coordinated colony of organisms, in the class Hydrozoa, that live together. They all belong to the genus *Physalia.* Their air bladder, also known as the pneumatophore, is filled with an unusual combination of gases (such as argon, xenon and nitrogen) that keeps the colony afloat but at the mercy of water and wind currents. Individual organisms making up the colony have specialized jobs. Some are concerned with reproduction and others with digestion of food, for example. Some of the feeding tentacles may reach more than 30 feet in length. The colonies themselves may sometimes reach more than 100 feet in length. Like all other organisms in the phylum Coelenterata, *Physalia* are equipped with trigger released stinging cells known as nematocysts. They use these structures to capture and kill prey items. Some species have toxins that are as potent as cobra venom.

Specialized Insect Predator and Prey

Cicadas are very common members of the insect order Homoptera. Most species grow up to two inches in length although some can easily double that number. These insects are more often heard than seen. They tend to stay well hidden among the leaves of trees and shrubs but the males emit loud, piercing buzz-like songs. In some parts of the South, the creatures are called "jar flies." A popular bit of folklore is that the insect song may be silenced by squeezing the trunk of the tree it is singing from. Of course, this story has no basis in fact. Among the many species, two major groups are recognized. The division is based on their life cycles. Annual cicadas complete their life cycle in only one year. Some periodical cicadas, on the other hand, require up to 20 years in order to metamorphose. They mature from small larvae that emerge from eggs laid in twigs, to nymphs that live underground, to adults. As the nymph moves above ground, it often leaves its exoskeleton attached to tree trunks when the adult emerges from it. Adult cicadas often fall victim to a rather large wasp known as the "cicada killer." The female wasps may actually attack a cicada in mid-flight and paralyze it with a sting. The wasp may then fly, or drag, the still-living cicada into its underground burrow where she lays eggs upon the helpless insect. When the larvae hatch, they feed on the living cicada.

Sponges Are Animals

Synthetic sponges are widely available today. However, at one time, the only sponges for bathing and cleaning were natural sponges. Sponges are the simplest known animals. They belong to a phylum of animals called Porifera that includes more than 4,000 named species. They inhabit a variety of freshwater and marine habitats. Two noteworthy examples of freshwater sponges include those of the genus *Ephydatia*, the river sponge; and *Euspongilla*, the pond sponge. The latter group is often green in color due to the presence of algae that live within their tissues. Sponges often reproduce sexually to generate swimming larvae. Once the tiny animal anchors itself to the ocean floor, or to some similar substrate, it remains there and matures. Sponges range in size from almost microscopic organisms to some that are bigger around and taller than a human. Sponges mostly feed on tiny microscopic organisms known as plankton. They do this by passing water through pores in their body wall and filtering out the organisms.

How Many Eyes and Legs Does a Spider Have?

Spiders have two major body regions, the cephalothorax (combining the head and chest) and the abdomen. All spiders have eight legs but the number of eyes may vary among species. Most spiders usually have eight eyes, or less commonly six, some pairs of which may be larger than the others. A few spider species are entirely blind due to the absence of eyes. The spiders with the best vision are, apparently, the jumping spiders in the family Salticidae.

Metamorphosis: A Changing Lifestyle

Metamorphosis is not unique to insects. Many organisms, in fact, have life cycles in which specific stages are vastly different from others. Usually the immature stage of an animal's life cycle looks and behaves very differently from the adult in species which metamorphose. Even among insects, there are different types of metamorphosis. The most complex form of insect metamorphosis is called holometabolic metamorphosis. It is commonly known as complete metamorphosis. When an ovum is fertilized by a sperm, the zygote first develops into a worm-like larva. Examples of larvae include maggots from flies, and caterpillars from butterflies and moths. Some beetle larvae live under ground and others under water. Larva may molt several times as they increase in size. Some scientists use the term "larval instar," or simply "instar," to refer to these periods. The next major stage of this form of metamorphosis is called the pupal stage. Pupae are mostly non-moving organisms, inside a protective covering, that are undergoing a dramatic reorganization of their tissues. The protective coverings of pupae are called "pupal cases" or puparia. The moth cocoon and the butterfly chrysalis are two examples of pupal cases. It is from these pupal cases that adult organisms emerge. Examples of insects that carry out this type of metamorphosis (other than butterflies, moths and flies) include beetles, wasps, ants and fleas. Paurometabolic metamorphosis and hemimetabolic metamorphosis patterns lack larval and pupal stages. In both cases, a wingless version of the adult emerges from the fertilized egg. On land, it is called a nymph and closely resembles the adult. In water, it is known as a naiad and may be very different from the adult. Grasshoppers produce nymphs and dragonflies produce naiads.

Starfish

Although their common name would imply so, starfish are not fish at all. Instead they belong to the phylum Echinodermata (this word means spiny skin) along with other animals like the sea urchin, sand dollar and sea cucumber. Starfish belong to the class Asterodia, a word meaning "star shaped." These organisms begin their life as a fertilized ovum develops to a tiny larva that slowly matures to the recognizable shape of the adult. Most starfish have five arms. Some species, however, may have more than 20. Starfish have an internal network of canals and tubes that enables them to shift air and water through their bodies in both directions. They use this water vascular system for movement and for feeding. Coupled with their very muscular tube feet, hundreds of which are found along their arms, starfish utilize pressure from their water vascular systems to pry open the shells of clams and other molluscs with two shells. Once the starfish form a small opening between the shells, they actually protrude their stomach into their prey and digest away the contents. Starfish are known for their incredible powers of regeneration. They can re-grow entire arms that were lost or damaged due to predation, accident or disease. In fact, some species may regenerate a complete starfish from as little as one arm and a portion of the central body region.

The Longest Animal?

The ribbon worm holds this record. More than 550 species of these worms, in the phylum Nemertea, have been described. Another common name is "proboscis worm." Some species are less than one inch long; most others are less than two feet long. Their common name is descriptive of their elongated and flattened body shape. Some nemerteans live on land but most occur in marine habitats. The worms feed on dead or dying animals. In the 1860s, the largest known specimen was documented to have a length of 180 feet. It was found in Scotland.

Camping Caterpillars

The Eastern tent caterpillar, *Malacosoma americana*, is often observed during spring in the Eastern United States. Several hundred caterpillars may cooperate to spin silk threads and build the tents that are so often seen on apple trees, cherry trees and other species of trees. The caterpillars enter and leave the safety of their silken tents as they feed. Sometimes, they may defoliate their host tree or other nearby trees. With time, the caterpillars metamorphose into Eastern tent caterpillar moths. These adult moths will mate and lay eggs on the twigs of host trees. The following spring, the cycle will begin again.

Ballooning of Spiders

An unusual method of movement has been observed in several spider species. Depending on the species, both adults and juveniles (known as spiderlings) may balloon. They use their powerful leg muscles to climb toward an area exposed to a gentle wind. The spiders may then release a silk line from their spinnerets and loosen their grip to be carried away by the breeze. Ballooning spiders have been recovered from ships nearly a thousand miles from shore.

Gills, Not Just Fish Have Them

When most of us hear the word "gill" we immediately think of the feathery organs, rich in blood vessels, that fish use to extract oxygen from the water. However, many invertebrate organisms also utilize these structures for gas exchange in water or with moist air as they move about on land. Just a few examples of the many animals having gills include squid, clams, oysters, lobsters, crabs and crayfish.

A Clear Case of Mistaken Identity

One of the most enduring myths in biology involves those organisms that are commonly called by names such as "granddaddy longlegs" or harvestman. They are arachnids and belong to the order Opiliones. A widely circulated story about these organisms is that they are one of the most poison spiders known, but are unable to kill humans because their mouth parts cannot penetrate human skin. Nothing could be further from the truth. In the first place, harvestmen are not spiders. Spiders belong to the order Araneae. Opiliones and Araneae share several similarities but have differently developed reproductive systems. Opiliones also lack the characteristic waist-like constriction found on spiders' bodies, between their heads and abdomens. Further, harvestmen do not make any sort of venom like spiders. Some species, in fact, feed mostly as scavengers although many hunt and prey upon tiny animals. The varied diet of harvestmen also includes fungi and plant materials. No spider is known to consume those things. Harvestmen lay their eggs within damp soil or similar substrates. As a general rule, spiders produce silken egg sacs. It may be that the myth about harvestmen being poison spiders developed by way of confusion about common names. There is a species of spider in the family Pholcidae (*Pholcus phalangioides*) that is called "granddaddy longlegs" in some parts of the country. Another common name for the organism is the "cellar spider." These creatures are spiders in every sense of the word. They usually hang upside down in their webs and they do produce venom. Cellar spiders have an unusual defensive habit of vibrating violently within their webs when disturbed. It has been suggested that this behavior may serve to deter or confuse a potential predator. Although *P. phalangioides* do capture prey in their own webs, they are also known to creep away from their webs on hunting missions. During these times, they enter the webs of other spiders, kill them and then devour them. After the meal, the cellar spiders return to their own webs.

Carpenter Bees

These large insects belong to the same order (Hymonoptera) and family (Apidae) as bumblebees. They are often mistaken for one another. Bumblebees are very hairy and are usually found near flowers. Most species live in colonies like honeybees but are often solitary while foraging for food. Carpenter bees, on the other hand, have no hair on their backs. They are usually observed in small numbers near wooden buildings, benches, tables or other structures made of soft wood. They are menacing because they bore holes, just smaller around than a dime, deep into these wooden structures. They lay eggs in these holes. Unfortunately these bees tend to congregate at favorite sites and may destroy or weaken wooden structures with their repeated work. They leave behind small saw dust piles beneath the cavities they excavate.

Butterfly and Moth Larvae: Who Would Have Guessed It?

The following organisms have gone by various misleading common names. They are all actually the caterpillar or larval stage of various species of butterflies and moths. They include the cornstalk borer, leaf roller, Mexican jumping bean worm, alfalfa web worm, corn worm, oak leaf miner, carpenter worm, bag worm, measuring worm or inchworm, army worm, leaf worm, cut worm, woolly worm, silkworm, oak worm, tobacco worm, tomato worm, hornworm, catalpa worm and many others.

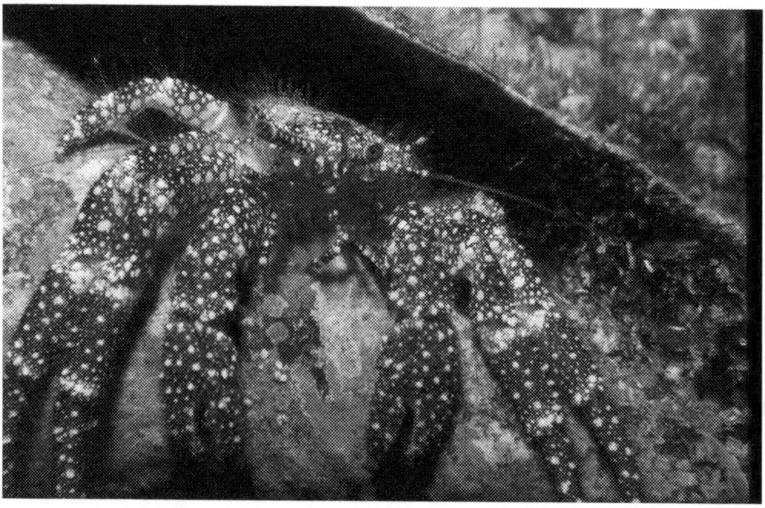

A hermit crab peering from its shell. Courtesy of NOAA/ National Marine Sanctuaries Media Library. Photographer not identified.

Crabby Hermits

Beach combers and pet owners all over the world are familiar with those animals we call hermit crabs. This common name is applied to many, many different species of crustaceans that live within abandoned seashells or similar homes. Some species are aquatic and others terrestrial. Throughout their life spans, most species of hermit crabs routinely molt their own exoskeletons as they mature, just as many insects do. As they grow, they must also secure new housing in the form of a seashell or other suitable covering. The hermit crab's body is modified to include only two legs for walking and large claw that acts as a door when they retreat into their shells. Other legs are modified for holding fast to the interior of the shell. Sea shells are not, however, the only homes inhabited by hermit crabs. Some species regularly utilize hollow lengths of bamboo or other light woody material. Some live in coconut shells. Occasionally they use man-made materials, such as cans, for their homes. One of the largest known land crabs, the robber crab (*Birgus latro*) is sometimes called a hermit crab due to the fact that it frequently utilizes seashells for protection. Robber crabs are said to be able to open coconuts. No credible documentation of this behavior exists.

Look At Those Bloody Worms

Unless you are a biologist or a fisherman, chances are strong that you have not encountered the interesting organism known as the bloodworm. Bloodworms are actually just worm-like in their appearance. They are the larval stages of some species of flies in the family Chironomidae. The adults are known as midges. They look a little like small, furry mosquitoes. Some midge species are biting pests but those that produce bloodworm larvae generally do not bite. They can pose a nuisance during their mating season. During warm weather, male midges crowd together in dense swarms above lakes and streams. When female midges are drawn toward the swarm, mating occurs in mid-flight. The females deposit their eggs on the surface of the water. The developing larvae display their brilliant red color due to the ability of their blood to store oxygen. The larvae build and live within tubes of mud, mucous and debris on the floor of a stream or pond. They seldom reach more than one-half inch in length and have tiny legs that give away their non-worm heritage. Once they break away from their pupal cases they emerge as adults on the surface of the water and then fly away. Adults have a very short life span, sometimes living only a few hours or days. Both adult midges and their larvae are a favorite food of many species of fresh water fishes. The bloodworm larvae are sometimes freeze-dried and sold as fish food and bait.

Dancing Honeybees

Honeybees of the species *Apis mellifera* have been kept in culture for decades. They have also been the subject of much biological research. Farmers value them for the honey they produce and for their ability to efficiently pollinate a number of farm crops. Worker bees leave the hive in search of food. Scientists have discovered that these worker bees communicate distance and location of particularly good food sources to one another by performing dance-like movements as they enter the hive. Depending on the distance of the food from the hive, one of two basic types of dances is performed. If a worker has found food within about 80 meters of the hive, she will perform what is called the round dance. She will enter the hive and regurgitate a small amount of food to allow the other workers to smell and taste the food she has found. As she does this, she will dance in a circular pattern within the hive. Instinctively, the other workers search for the food source but tend to limit their foray to an 80 meter radius surrounding the hive. When food has been found further away, the worker that located the food will perform a more complex ceremony known as the waggle dance. She will again allow others to taste and smell the food she has found. She will also provide her fellow workers with a direction, based on the position of the sun, by walking within the hive. By shaking her abdomen at a particular rate, the female conveys information about the distance of the food source from the hive. Scientists have recently discovered that bees also appear to communicate information about the height of the food source from the ground during their waggle dances.

Earthworms

Several species of earthworms exist through the world. They mostly feed on debris at and below the soil surface. As they digest and extract nutrients from their food, earthworms excrete a rich mixture of feces and soil called castings. In the 1800s the naturalist Charles Darwin estimated that each acre of farmland in his native England contained more than 50,000 earthworms. Earthworms have a primitive brain and most species can detect light and dark. They do so by way of photoreceptors which are embedded in their skin. The worms always seek out areas of darkness. Their skin is covered with tiny bristles, called setae, and has mucous for lubrication. The bristles help the earthworms move through soil. Earthworms are hermaphroditic, meaning they have both male and female sex organs. However they reproduce sexually, with both members of the mating pair exchanging sperm. A worm's sexual organs are contained within the smooth band of skin located about a third of the way from its mouth. The region is called the clitellum and is the site most often used by fisherman to secure bait worms to a fishing hook. Although earthworms can re-grow parts of their body that have been damaged or lost, this power is limited. Earthworms can bleed to death if cut in two. Contrary to widely held opinion, one cannot cut an earthworm in half and expect it to re-grow lost parts to form two new worms. The largest known species of earthworm is the giant earthworm, *Megascolides australia*, found in Australia. Adult worms of this species may reach approximately 10 feet or more in length.

Aquatic Bears

Several genera and species of tiny animals are called "water bears" or tardigrades. Some biologists classify them into phylum Arthropoda, along with insects, crabs and spiders. Others assign them to their own phylum, Tardigrada. They range in size from microscopic to just over one millimeter in length. These animals, with their segmented bodies, four legs and claw-like appendages, may remind one with a good imagination of a tiny bear. Water bears feed on various plant, animal and protozoan organisms. They occur in a wide variety of habitats such as soil and moss. Some species live in polar ice caps and others in hot springs. A few species may even be found in rain gutters. Most of these animals inhabit temperate environments. Some species of tardigrades can dry and encyst during hostile ecological conditions. They survive in this state and become active during favorable circumstances. There are some reports of water bears that have become active from within decades-old, dried moss specimens in herbarium collections.

Moths with a Taste for Blood

A few species of moths in the genus *Calpe* are known to routinely feed on the blood of mammals. They do not appear (according to most biologists) to be entirely capable of making wounds in the skins of mammals in order to drink blood. Rather, they seem to take advantage of existing wounds. These moths extend their mouth parts into the wound and suck blood with their proboscis. No known species exists entirely on blood for a food source.

Solitary Bees and Wasps

Many species in the insect order Hymenoptera live in complex social colonies. Examples include the honey bee, most species of ants, paper wasps, yellow jackets and countless others. However there are an impressive number of species of hymenopterans that do not live in colonies. Instead, they produce individual nests and spend most of their lives foraging for food on their own. Examples include the digger wasp, mud dauber (also called the dirt dauber) and the potter wasp. They are collectively called the solitary wasps and solitary bees.

Bird Eating Spiders

Including their legs and bodies, these large spiders of South America may reach the size of a dinner plate. They are able to catch small mammals and birds but do so by actively hunting them, not by capturing them in webs. Most often they feed on smaller organisms such as insects. Some species tend to live in burrows in the ground. The bite of bird eating spiders may be painful but it is generally harmless to humans.

Where Do Pearls Really Come From?

Throughout history pearls have been highly prized for jewelry, buttons and various other types of ornamentation. Pearls occur naturally in a variety of freshwater and saltwater molluscs including clams, mussels and oysters. The process begins as an irritating substance (such as a grain of sand, or a bit of shell or other debris) gets trapped within the animal's tissues. This causes irritation, and the organism slowly begins to cover the foreign particle in nacreous or pearly layers of chemicals. Pearls vary widely in their color, shape and size. White, pink, yellow, black and blue pearls have been recovered. The most commercially sought after pearls are usually found in oysters of the genus *Pinctada*. Several years ago, farmers developed a method to artificially produce pearls. They mechanically insert an irritating particle and/or a blank pearl-like particle, and the mollusc does the rest. Another product of these and other molluscs that was once highly sought after was called "mother of pearl." This substance is found on the interior lining of the shells covering the bodies of these organisms.

A World Full of Insects

For the most part, biologists regard insects as the most successful group of organisms on the planet today. They outnumber humans by an estimated ratio of 200 million to one. More than 75% of the known animal species are members of class Insecta, also called class Hexapoda. The group is extremely varied. The individual organisms move by walking, swimming, flying, burrowing or jumping. Different insects produce deafening sounds, deadly stings or carry fatal diseases. Others pollinate crops, inspire songwriters and help scientists cure illness. Some species have life spans of merely a few hours while others may live for years. They range in size from practically microscopic to well more than one foot in length. Some insects are able to withstand extreme environmental pressures. A species of beetle, *Niptus hololeucus*, lives successfully on cayenne pepper and as even been observed living in cork stoppers on top of bottles of cyanide. More than 750,000 other species of insects have been named. Biologists estimate that at least as many more remain undiscovered. Some place the estimate of unnamed insect species above seven million.

Tubifex Worms

If you keep an aquarium, chances are that you have purchased a pet food called "tubifex worms." They are often sold, freeze-dried, in small cubes and can be found in most any pet supply store. These unusual creatures, *Tubifex tubifex* and similar species, belong to the family Naididae. They are grouped within the class Oligochaeta, along with earthworms and their relatives. Tubiflex worms are often found in fresh water lakes and streams. They swallow mud and extract nutrition from it as it passes through their digestive systems. It is of note that these worms spend most of their time suspended, head first, within a tube of mud and silt. This may account for another common name for the species, sludge worms. Some specimens may exceed 10 inches in length.

Octopus Versus Squid

Both the octopus and squid belong to the phylum Mollusca and class Cephalopoda. Many species of each are known. They share numerous similarities. For example, all known species of squids and octopuses live in marine habitats and breathe by gills. They all have similar circulatory and digestive systems. Both have a tough beak for feeding and can move in rapid bursts of speed by way of a water-based jet propulsion system. Both may release clouds of pigment, commonly called ink, to confuse predators. Also, most species of squids and octopuses have skin cells called chromatophores that are rich in pigment and allow for quick color changes to blend with the environment. They also serve to flash warning color patterns. How are the organisms different, then? Octopuses, as the name implies, have eight arms. Squids have ten, with two longer arms being modified with tentacles for gripping prey. Most octopuses have toxic saliva while squids do not. Squids move by swimming. While octopuses are excellent swimmers, they most often crawl along the ocean floor and remain in seclusion except when feeding.

Mosquitoes

Mosquitoes are members of the insect order Diptera and are close relatives of the house fly. They reproduce very quickly and may take as little as three weeks to complete their life cycle. After adult mosquitoes mate, the fertilized eggs are usually deposited in water. In most species, the larvae float near the surface of the water so that they may exchange oxygen and carbon dioxide. Some species have larvae that actively hunt for small prey items beneath the water. Adults will emerge from pupal cases and begin their life on land. Most species of mosquitoes feed on flower nectar and/or other juices of plants. In those species that do drink blood, only the female does so and only just prior to mating. The blood meal apparently provides her with extra nutrition for egg production. It is of note that many species of mosquitoes (but certainly not all) are vectors for bacteria, protists and other pathogens that cause a variety of human diseases. In fact, more human deaths involve mosquitoes than any other animal.

Winged termites like these leave the parent colony to form a new colony. Courtesy of USDA-ARS. Photograph by Scott Bauer.

Termites

Termites, also called "white ants" are social insects found in many parts of the world. They belong to the order Isoptera. True ants are in the order Hymenoptera. Each termite colony usually has just one female, the queen, which is capable of reproduction. Sometimes, multiple queens are found within a single colony. The queen secretes a chemical that renders the other females in the colony sterile. She may lay up to one thousand eggs per day. With her massive abdomen swollen with ova, she is often much larger than others in the colony. The queen termite becomes so large, in fact, that she is essentially helpless without the care and feeding of her workers. When the colony grows to be very large, winged males and females leave the parent colony to mate with members of other colonies and thus begin new ones. During these migrations, people often mistake the termites for wing-bearing ants.

Not a True Spider, a False Spider

One common name for a particular group of unusual animals is false spider. Another name is wind scorpion. These organisms are related to spiders and scorpions in that they share the same phylum, Arthropoda, and the same class, Arachnida. The things that make these creatures unique, leads scientists to place them in a separate order called Solifugae. False spiders are carnivorous and have a mighty appetite. They are active mostly at night in their tropical to semitropical habitats. Many species are covered with tiny hairs. Also, false spiders lack spinnerets. These are the organs at the most posterior end of the abdomen that spiders use to spin and manipulate their silk threads.

Largest Insect

In terms of absolute body mass, the beetle *Goliathus regius* is the largest known insect. Other species in the genus come close but most biologists agree that *G.* regius (commonly called the "goliath beetle") wins the honor. The goliath beetle belongs to the beetle family known as Scarabaeidae, commonly called scarabs or "scarab beetles." Like all insects, *G. regius* have six legs. They can reach about four and one-half inches in length and more than 2 inches in width. Their bodies are patterned with bands of black, white and brown. They mostly live in rainforests on the continent of Africa. Like other members of the order Coleoptera, goliath beetles undergo a complete metamorphosis. In fact, their larvae may reach up to six inches in length if they are well nourished.

The Coffin Fly

The coffin fly, *Conicera tibialis*, is a relative of the housefly. It has a strange and poorly understood life cycle. The organism's larvae, called maggots, feed on the remains of dead animals. The adults are often found above ground. However, laboratory experiments and field observations seem to support the conclusion that the coffin fly can complete its entire life cycle underground. In other words, some flies may never venture above ground for any reason. The coffin fly earned its common name due to extensive reports of the maggots being found on buried bodies. Dead bodies may be infested with eggs and larvae before burial.

Round Worms

Officially, round worms are simple organisms that the lack true body segmentation seen in earthworms and leeches. As a group, they are often classified into a single phylum called Nematoda or Aschelmenthes. In some classification schemes, Aschelmenthes is designated as a phylum and Nematoda a class. Another common name for the group is eelworm. More than 75,000 species are recognized. The roundworms are a tremendously varied group in terms of dimension and lifestyle. They range in size from microscopic, to one species that may routinely exceed 20 feet in length. Free-living nematodes may be found in water and in soil. They are valuable to the ecosystem in that they help to enrich the soil and provide food for other organisms. Some species of roundworms are parasites and may infect a wide variety of animals, including humans. Animals as diverse as beetles, caterpillars, dogs and termites may be hosts for the worms. A few species of round worms parasitize plants. Root-rot nematodes are an example. They block passage of water into plant stems and may cause death in severely infected plants.

About Bumblebees

Bumblebees belong to the insect order Hymenoptera, along with ants, wasps and many other relatives. True bumblebees of the genus *Bombus* are social insects. Bumblebee colonies are most often found in underground nests, complete with queens, sterile workers and, rarely, male drones. They make very small quantities of honey. Bumblebees produce sounds from their wings, and a characteristic humming tone from their respiratory system. The old story about bumblebees being aeronautically incapable of flying is not true. The story has been widely spread in the popular press and was based on earlier misunderstandings about bumblebee wing structure. A couple of other insects are often called bumblebees by mistake. One species in the genus *Psithyrus* can actually take over a *Bombus* colony and force the workers to feed and care for its offspring in areas where their ranges overlap.

Slugs of the Sea

The common name slug is usually reserved for those land organisms that are often described as being "a snail without a shell." Most everyone is familiar with these creatures that frequent yards and gardens, leaving a glistening trail of slime in their pathways. They belong to the phylum Mollusca along with the octopus, squid, clam, oyster, snail and other organisms. The order Stylommatophora includes land snails and land slugs. Land slugs may range in length from less than one inch to over 10 inches. Color depends mainly on the species but includes gray, black, brown, white, yellow and orange. Most land slugs have a tiny, rudimentary shell that goes unnoticed by the average person. The slugs feed primarily on plant material and a few species are garden pests. Some eat dead animal remains and/or feces. A close relative of the land slugs, also within the phylum Mollusca, is often called the sea slug. Sea slugs belong to the order Nudibranchia. Several genera and species are known. They live in the seas and breathe through masses of gills on their backs that superficially resemble a spray of tiny flowers. Like many of their land-living relatives, sea slugs have no shell.

CHAPTER FOUR

Kingdom Animalia: The Vertebrates

Migrating Fishes

Some fishes actually travel long distances to their breeding grounds. Generally the migratory fish are divided into two categories. Those that live in salt water as adults but return to freshwater to breed are called anadromous fish. The salmon is the classic example. Not all species of salmon are migratory, however. Lampreys are another group of anadromous migrators. Fishes that live in fresh water as adults but return to saltwater to breed care called catadromous fish. Freshwater eels in the family Anguillidae are good examples. All known species may travel many miles to breed in the Sargasso Sea. Only one freshwater eel, *Anguilla rostrata*, is found in the United States. Striped bass are also catadromous in their migratory routes.

Senses of Snakes

Snakes have no external ears so they are not able to hear in the traditional sense. Yet, they do have internal hearing mechanisms. So what about the classic image of the snake and snake charmer? Cobras very likely dance in response to movement of the charmer and in reaction to vibrations which they sense through nerves in their skin. Most snakes have fairly well developed senses of taste. Many snakes can see well, except during molting, when the skin covering their eyes becomes clouded. Some snakes, called "pit vipers" sense heat and likely interpret a visual image based on body warmth of potential prey items. Examples of pit vipers include rattlesnakes and copperheads. These snakes give birth to live young that produce the same deadly venom as their parents.

Pink Flamingos

Those interesting plastic birds that some people place in their yards for decoration are actually modeled after real birds. Flamingos belong to the family Phoenicopteridae and include species that approach 50 inches in height. Flamingos occur in Central and South America, Asia, Africa and elsewhere. *Phoenicopterus ruber* is found in the United States. These long-legged birds, with their curved bills, vary from white to very dark pink or red in color. The color is not genetically based. Flamingos that eat large amounts of red to pink colored materials (such as algae, insects and shrimp) often pick up those same pigments in their feathers. The feathers of even a very dark red flamingo will eventually pale and become white unless the bird continues to consume foods rich in the pigment carotene. Flamingos enjoy a widely varied diet which also includes fish and other organisms that do not contain carotene.

Bird Hygiene

Most species of birds are meticulously clean. They spend a great deal of time keeping their feathers and skin in good order, removing parasites and keeping up appearances in other ways. Most birds have an oil gland located at the base of their tail. They will wipe their beak across the gland and spread the oil on their feathers in an act called preening. The oil helps keep the beak and feathers healthy and waterproof. Some species of birds engage in an unusual behavior called anting. In this activity, the birds kill ants with their beaks and rub the secretions over their skin and feathers. Some scientists believe this behavior reduces the presence of parasites. As a final point, most all birds enjoy a water bath or dust bath very frequently.

Brook trout. Courtesy of United States Fish and Wildlife Service (USFWS). Photograph by Eric Engbretson.

Jewel of the Southern Appalachians

In some remote, cold, high elevation streams in the Southern Appalachian Mountains there are a few remaining populations of *Salvelinus fontinalis.* Local citizens call this fish the "brook trout" but it is actually more closely allied with the fish known as char. This native species was probably once very common. Their preference for pure, clean water is not in harmony with the logging, building and damming activities of recent times. These actions contributed to the decline of the brook trout's numbers. The fish rarely exceed one foot in length. They are a favorite of some fishermen and are noted for their taste as a game fish. Continued population declines in recent years have led to restrictions on fishing and capture of this interesting fish.

Largest Known Animal

The largest known animal on the planet today is the blue whale, *Sibbaldus musculus.* In fact, most scientists regard these gigantic mammals of the family Balaenoptera as the largest animals in the entire history of the Earth. The average male is between 80 and 90 feet long and weighs 100 tons. Some have reportedly exceeded 100 feet in length. They are slate blue in color and have hearts the size of small automobiles. It has been said that a child could swim comfortably through the largest blood vessel in the massive whales' bodies. Females give birth to young, called calves, which may be up to 25 feet in length and 50 tons in weight. Despite their massive size blue whales belong to a group of whales that feed primarily on some of the smallest animals, called krill. From the days of whaling we know that their stomachs may contain well in excess of one ton of food at any given moment. Blue whales gulp in large amounts of water and filter food from it. They do so by expelling the water through modified hairs called "baleen plates" that hang at the opening of their mouths. They sometimes take larger prey. Like other whales and related aquatic mammals, the nostrils of blue whales are found on the backs of their heads. This allows the animals to take in and expel great breaths of air as they reach the surface of the water. Blue whales also hold the distinction of making the loudest animal sounds on record. Some of their vocalizations may approach 190 decibels in intensity.

Some Bats Are Real Vampires

Three known species of bats have been observed to regularly feed on the blood of other animals. Blood is, in fact, their primary food source. It is important to note that these organisms have a very restricted range near Chile, Mexico and Argentina. No blood-eating bats are regularly found in the United States. The most common species is *Desmodus rotundus.* Two other species, *Diphylla ecaudata* and *Diaemus youngi* share portions of their ranges with *D. rotundus.* Vampire bats feed by cover of darkness as a general rule. They have been known to prey upon farm animals such as cows, horses, chickens and pigs. They also drink blood (a practice called hematophagy) from wild deer, squirrels, birds and any number of other animals. The bats bite the skin of their host in order to feed. Their saliva contains chemicals to discourage blood clotting. They lap blood from the wounds with their tongues. Unless they are disturbed, vampire bats will drink blood until their abdomens become swollen. In a single night, one bat could almost never consume more than about two teaspoons or so of blood. It would take multiple attacks upon a single large animal, over the course of several days, to cause blood loss severe enough to kill. Sometimes this does happen. Like so many other species of bats, vampires have well developed senses of smell and vision. They can fly and walk. They often rest in groups, in hollow trees, during daylight hours. Humans rarely fall victim to a vampire bat's appetite for blood. They usually sleep within well protected shelters and have the dexterity to fend off a biting bat.

Living Dragons

In 1912, *Varanus komodoensis*, more commonly called the Komodo dragon, was documented by mainstream scientists as the largest known living lizard. It shares its common name with Komodo Island, the largest of a small chain of islands to where it is restricted. Adult males may reach up to ten feet or more in length. Some weigh more than 300 pounds. The organisms feed mostly on dead organic material known as carrion. However, they do routinely and actively hunt large prey items. Because their mouths are rampant with bacteria and other microorganisms, a bite from the Komodo dragon usually means a lingering (sometimes days-long) death for prey. The dragons follow their prey until the unfortunate victims succumb to infection. Some biologists believe the reptile may make its own toxins as well. Even before their discovery by science, tales of humans falling victim to Komodo dragon bites were in circulation among people on nearby islands. Scientists regard many of these stories as highly credible.

Two Old Buzzards

In the Southeastern United States, there are two species of birds that are often called by the common name buzzard. Biologists prefer to call them the turkey vulture, *Cathartes aura,* and the black vulture, *C. atratus.* The turkey buzzard is the larger of the two species. Its average length is about 30 inches, with a wing spread of six feet. The slightly smaller black buzzard is about 27 inches long, with a wing spread of about four and one-half feet. The mature turkey buzzard has a patch of red, featherless skin on its head. Juvenile turkey vultures may be mistaken for black buzzards. They may sometimes take small, live prey items. However, both species appear to prefer dead and decaying animal remains. This accounts for another common name often applied to both birds, "carrion crow."

Bird Eggs

Bird eggs, or ova, are among the largest known cells. These specialized cells are for reproduction. The largest known bird egg (and the largest known cell) belongs to the ostrich. They are nearly seven inches long. Some extinct birds are believed to have produced even larger eggs. Hummingbirds hold the record for the smallest bird eggs. Some measure less than one-half inch in length. Most species of birds have only one functional ovary, the site where eggs form and mature. The developing eggs travel through the oviduct from the ovary. During this portion of the journey, the eggs may be fertilized if the female bird has engaged in breeding. Calcium-rich protective shells are formed around the softer material of the egg in a structure called the uterus. The eggs are then released through the cloaca, the common opening for egg-laying, waste disposal and breeding.

Rapid Heart Rate

Some species of hummingbirds have a heart rate which easily exceeds 1000 beats per minute during intense activity. Compare this with the average resting heart rate of a healthy human, 70 beats per minute. Hummingbirds have an incredibly rapid metabolic rate and require much food and oxygen to keep pace with their energy demands. Most species have a life span of two or three years although some can live more than 10 years.

It is Reining, Deer

Those organisms commonly called reindeer belong to the genus *Rangifer*. Species are found native in Europe, Asia and regions of the Arctic. The common name of these organisms reflects the fact that they were often bound with reins and used to pull sleds. They have been used for this purpose for centuries. Their large hooves are ideally suited for travel in deep snow. Both sexes normally produce antlers, an unusual phenomenon. The organisms may utilize their antlers and their hooves to dig for plant material beneath the snow.

Largest Known Fish

The largest fish known to science is called the "whale shark," *Rhineodon typicus*. The fish may sometimes reach lengths of more than 70 feet but one measuring 40 feet is regarded as being especially long. Humans rarely get a glimpse of these massive creatures. Mature adults have a mouth that often opens more than four feet across. Mouths of whale sharks are equipped with thousands of tiny teeth. Whale sharks are considered to be harmless to humans however. They feed mostly on microscopic organisms and small fish.

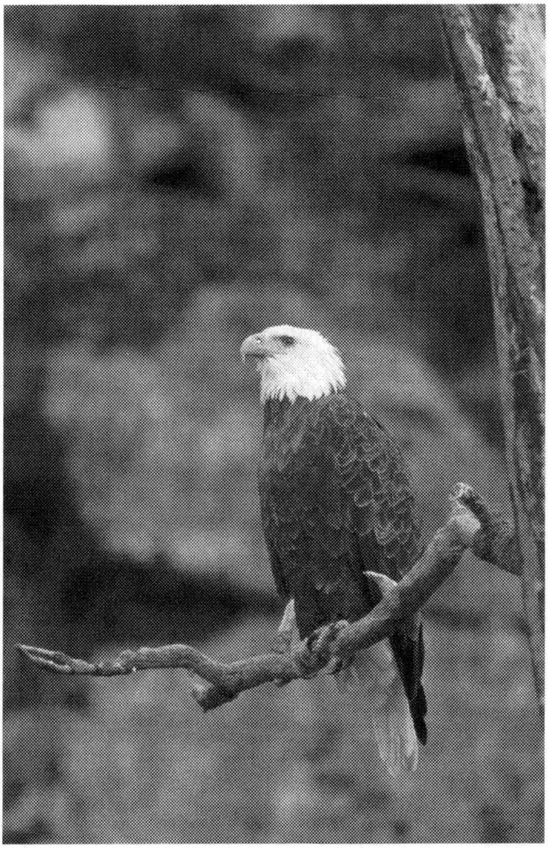

The bald eagle. Courtesy of USFWS. Photograph by Gary Kramer.

White Head

The term leucocephalus literally means "white head." It is part of the scientific name *Haliaetus leucocephalus*, an organism more commonly known as the bald eagle or American eagle. This huge bird of prey is, of course, pictured on the national emblem of the United States. Many years ago the term bald was not used to describe a hairless person, but rather a person with white hair. If you called someone bald, that meant that their hair had turned white. The distinctive white feathers begin to appear on the adult eagle's head by the time it reaches about seven years of age. Juvenile birds have gray head feathers. Adult birds may be up to three feet tall and have an eight foot wingspan. Bald eagles mate for life and have a strong tendency to reuse the same nest, year after year. They are often found near water. The birds feed mostly on fish which they capture while wading or by sweeping them from near the surface of the water. Their diet, however, is varied. They may catch and eat a variety of mammals up to the size of foxes and rabbits. They are sometimes observed catching other birds in flight and may sometimes even snatch food from other large birds of prey. They are also known to feed on snakes and carrion. By all indications, bald eagles have outstanding eyesight.

Constricting Snakes

Several species of snakes are called constrictors because they do not utilize venom to kill their prey but, instead, suffocate their prey by wrapping coils of their body tightly around their victims. These snakes may begin their predatory activities by biting their prey in an effort to startle or disorient them. They quickly wrap their bodies around the prey item and slowly suffocate it to death. As the prey exhales, the snake tightens its grip until the animal can no longer breathe. Constrictors swallow their prey whole. Their saliva is rich in mucous and digestive enzymes. The snakes have the ability to unhinge their jaws, and have very strong skull bones to prevent damage to their brains as they swallow. The largest and best known constrictors are probably large species of anacondas and pythons. Throughout history, accounts of giant constrictors have been circulated. Some reports make claims that these snakes reach up to 150 feet in length. Mainstream biologists do not regard these stories as credible. A 20 foot long anaconda is considered large. Some biologists accept the possibility that they may approach 40 feet in length. Pythons probably do not exceed 30 feet in length very often, if at all. Anacondas and pythons can attack, kill and swallow some surprisingly large prey items. Leopards, antelopes and gazelles are included in this well documented list. When snakes eat items this large, movement becomes slow and difficult making the animals more at risk for harm. A large meal may last some snakes for months before they need to hunt again.

Learning to Sing

How is it that a bird of a particular species is able to sing the correct song? The answer is not as straightforward as one would hope. It is true that most species of birds have particular and characteristic calls and songs that appear to serve a wide variety of purposes. However, the answer to our question depends largely on the species of bird. Some species are born with the instinctive ability to sing their group's songs. They will do so even if raised in isolation from other members of their species. Courtship songs of male cowbirds (*Molothrus ater*) are an example of an unlearned song. Other bird species seem to have the need to hear songs from members of their own kind. They are able to learn the songs as they are heard. It is of note that some species appear to be able to learn only songs of their species. Juvenile white-crowned sparrows (*Zonotrichia leucophrys*) seem to have a critical period during which they must hear other white-crowned sparrows sing, and during which they must practice their official song. If this critical period is missed, the birds can vocalize but are unable to sing in the correct fashion. Other birds, such as the mocking bird and blue jay, display extraordinary abilities to imitate a wide variety of bird songs and non-bird sounds. Most bird vocalizations arise from their song organ, called the syrinx. It is analogous to our own voice box or larynx but more complex. Some birds may make sounds in other ways. For example, the whistling sound that is often heard when mourning doves (*Zenaida macroura*) take flight is actually made by their wings.

The Sloth

Sloths are mammals with unusually slow metabolic rates. They are natives of the South American rainforests. Typically the organisms may be observed hanging upside down on a tree branch. Their fur is often colored green due to the presence of green algae that grow on the animal's hair. Two groups of sloths are recognized. There is a genus called *Choloepus* with two toes on the front legs; and another genus, called *Bradypus*, with three toes. Within these two genera, biologists typically recognize up to six species. Sloths are usually active only at night, but even then their movements are slow and calculated. They are herbivores, meaning that they eat only plants. Sloths very rarely leave the safety of trees. They do make regular trips to the ground to defecate.

Monkeys and Apes

Monkeys, apes, marmosets and humans all belong to the mammalian order Primates. In taxonomy the term is pronounced "pry-MAY-tees." As a group, the primates are noted for their good eyesight and well developed brain. The terms ape and monkey are often used interchangeably in error. Marmosets are about the size of a squirrel. Monkeys are substantially sized primates that often have long tails. A convenient way to group the many species of living monkeys is by geography. Old world monkeys are those that are found in Africa and Asia. They have a narrow nose pad with downward facing nostrils. Their tails are usually not capable of grasping. New world monkeys are known only from South America and Central America. They have flat nose pads with nostrils positioned sideways. Their tails are often capable of grasping branches to assist in climbing. Apes, in contrast, do not have a tail. Like old world monkeys, apes are known today only from Africa and Asia. Gorillas are the largest known apes. Other apes include the orangutan, gibbon and chimpanzee.

A Mole with a Pocket

There is indeed a marsupial mole in Australia. Female *Notoryctes typhlops* have a pouch for protecting their young just like kangaroos, opossums and other marsupials. The marsupial mole is approximately the same size as various species of moles found in yards across the United States. They live much the same lifestyle, burrowing beneath the surface of the ground and feed on similar types of food.

Browsers and Grazers

Animals that feed mostly to exclusively on plants are called herbivores. Organisms like cows, horses and zebras that feed mostly on grasses and other ground level plants are called grazers. Those that eat the leaves and stems of shrubs and other plants are called browsers. Deer and elephants are examples. Many herbivorous organisms feed on both types of plants.

Sea Otters and Tool Use

The sea otter is often observed floating on its back in the water as it uses a flat stone to help it feed on the mussels and other shell fish that make up a large percentage of its diet. These mammals usually rest the stone on their chest or belly and break the shells of their prey items open on the stones.

Was that a Turtle or a Tortoise?

These words have little to no significance in biology. They are used mostly in the common names of those organisms generally known as turtles. Throughout history, the word tortoise has often been applied to land-living turtles. Those associated with water are very often referred to simply as turtles. Occasionally the term terrapin has been used for both land dwelling and aquatic organisms. More often than not, it has been applied to many of the numerous species of turtles that humans have consumed for food. Turtles are reptiles that belong to the taxonomic order Chelonia. They are the only known reptiles with shells. The shells are actually derived from specialized skin tissues. The portion of the turtle shell that covers the back is known as the carapace. The segment covering the belly of the animal is called the plastron. These shells offer turtles varying degrees of protection. Turtle shells vary from being extremely hard to rather soft and pliable. Some of the animals can completely retract their legs and head in the shells. Others cannot. A few species of turtles (called soft shells) have only a tough, leathery covering that superficially resembles the harder shells of their relatives. Some turtles in fresh water, some in salt water, and others are strictly land dwelling. Some eat only plants while others are active hunters. Many species are known and the individuals range in size from only a few inches in length to more than seven and one-half feet. The largest turtle, *Dermochelys coriacea*, is known as the leatherback turtle.

Florida Manatee. Courtesy of USFWS. Photograph by Keith Ramos.

Animal Sirens

All manatees belong to the mammalian order Sirenia. This order takes its name from Greek mythology. The story goes that creatures known as sirens would sing such lovely songs that they would draw sailors toward their island. The seamen would lose control of their boats and crash on the rocks as they made their way toward the song. Some early explorers who saw manatees for the first time believed that they were seeing mermaids. Mermaids are often confused with sirens. In mythology, sirens had the body of a bird. Mermaids had the body of a fish. Today, manatees (Genus *Trichechus*) are endangered. A few may reach lengths of 12 feet and weight more than one and one-half tons. Most are significantly smaller. They tend to be very slow and shy creatures. They feed exclusively on plants. As their teeth become worn or damaged, other teeth from the back of the mouth move forward to take their place.

Disappearing Frogs

It is not an uncommon thing for frogs to burrow into soft mud to escape cold, heat and other environmental extremes. The continent of Australia has a group of several frog species that are so well-known for this feat, that they are often called "burrowing frogs." What makes them so noteworthy is that some spend most of their lives underground, in a state of dormancy. Adult burrowing frogs mate only during the rainy seasons. In some Australian desert environments, heavy rains may be years apart. After mating, the adults drink large quantities of water and store it primarily in loose layers of skin surrounding their bodies. The waterlogged amphibians then dig their way into dirt, sand or mud. They survive off of their stored water reserves and emerge only during the next period of heavy rain. Native inhabitants of Australia sometimes dug the frogs out of mud and drank the stored water from them. The largest species of burrowing frog known is *Heleioporus australiacus*. Other species in the genera *Neobatrachus* and *Limnodynastes* also store water and burrow.

Bird Nesting Behavior

Numerous species of birds are known to biologists. All birds deposit fertilized eggs into the environment. The developing young derive their nutrition from the yolk sac within the egg. Hatching time varies widely and may range from less than two weeks to more than 50 days. In some species of birds only one parent tends the nest. Other species divide the duty between parents and a few sometimes have the help of other adult birds. Some species of birds provide absolutely no care to their eggs or young but lay their eggs in the nests of other species. Such behavior is known as "brood parasitism." Newly hatched birds vary in their maturity and level of dependence on adults. Precocial birds can feed almost as soon as they hatch but rely on their parents for protection, and for modeling of appropriate behaviors. Examples include ducks, chickens and turkeys. Altricial birds are born very helpless and must be fed by an adult for several days to several weeks. Robins, sparrows and wrens are examples. Some species of birds have offspring within a single nest that hatch at different times. This practice is known as incremental brooding or asynchronous hatching. Owls, crows, storks and hawks carry out this process.

Their oldest and youngest chicks may be separated in age by several days within a single brood.

The Devil Dog

"Devil dog" is an obscure common name sometimes applied to one of the few species of giant salamanders found in the world. Hellbender is the more common vernacular name for *Cryptobranchus alleganiensis*. These organisms are found in freshwater habitats of the Eastern United States. They have flat, warty heads and compressed tails. Hellbenders reach an average length of 16 inches, although a few extreme cases in excess of 25 inches have been documented. The organisms feed mostly on insects, worms and crayfish. During mating, males prepare a nest-like depression for egg laying. They guard the eggs which are produced by the female in rope-like masses.

This Bird is Such a Lyre

One of the most beautiful and unusual birds, in terms of plumage display, is the lyre bird. The birds belong to the genus *Menura* and are known only from Australia. The showy tail feathers of the male lyre birds resemble the ancient musical instruments of the same name. The birds are the size of an average domesticated chicken and rarely fly. They live mostly on the ground. Other things that make these birds worth seeing and hearing are their remarkable powers of mimicry. It is said that they can imitate more sounds that most any other bird. In addition to being able to copy the songs of many other birds lyres can sound remarkably like chain saws, ringing telephones, computerized games and even speaking humans.

Boneless and Bony Fishes

When speaking of one species or type, use the word "fish." When speaking of two or more species or types, use the word "fishes." Many species of fishes exist throughout the waters of the world. All fish belong to the phylum Chordata and subphylum Vertebrata. However, one large group of fish in the class

Chondrichthyes is entirely devoid of bone. They do have an internal skeleton made of tough cartilage, the same general type of tissue making up the human nose and ear. Members of this group include the sharks, rays and skates. Other groups of fish have an internal skeleton made of both cartilage and bone. These fish belong to the class Osteichthyes. Examples include trout, bass and carp.

Budgies

Each year countless numbers of small birds called budgerigars or budgies (*Melopsittacus undulatus*) are sold in the United States and other countries as pets. They belong to the parrot family, Psittacidae. In the United States they are often called parakeets. That common name is applied to several members of the family. In their native Australia, budgerigars are often called betcherrygah. Substantial wild populations of mostly green birds still exist there. In captivity various other colors including white, yellow and blue are often prized by potential pet owners. Like their domesticated cousins, wild budgerigars feed mostly on seeds.

Do Cows Really Have Four Stomachs?

Well technically no, but sort of. Some people call four of the specialized parts of a cow's digestive system "stomachs" but cows really have just one. Cows and other mammals that have these sorts of complicated digestive systems are referred to by biologists as ruminants. Digestion of food by them is known as rumination. Animals such as these feed heavily on grasses and many other types of plant materials that are especially rich in a sugar called cellulose. No known animal can digest it. So, cows use the assistance of microbes to help them access the nutrition in these types of foods. When they graze, cows send a mouth full of roughly chewed food down their esophagus and into a holding tank called the rumen. It is especially rich in heat, gases, fluids, bacteria and other microorganisms. The microbes begin the work of chemically breaking down the fiber-rich food materials. From there the food passes into a second chamber called the reticulum. It also contains many bacteria and other organisms that help break down the plant fibers. Most often cattle will stop swallowing new food and concentrate on processing the already ingested food. They vomit the larger remains of their meal back into the mouth to be re-chewed, in a characteristic side to side motion, and swallowed again. This partially digested mass of food and liquid is called cud. You may sometimes observe a mass of the material being brought back up to the mouth by noticing a bulge rising up the animal's throat. The process continues until the food is mostly liquefied. It then passes from the reticulum into the omasum. It is a rather muscular portion of the digestive system from which water and some fats are absorbed. The omasum pushes food into the cow's true stomach, the abomasum. Here, many of the cow's own digestive enzymes continue the breakdown of the sugars. Most absorption of nutrients, and additional chemical changes, occurs in the animal's small intestine. Solid wastes are released from the large intestine. Other ruminants include sheep, goats, deer, elk, antelope, llama, giraffes, musk oxen, bison, buffalo and caribou. Camels and their close relatives have a similar but less complex digestive system and are not considered to be true ruminants.

When Foxes Fly

Pigs will never fly but foxes may. It all depends on which common name you use. The term "flying fox" has been assigned to several genera and species of very large bats that exist in various locales. Populations are known from Madagascar, Australia and India. Some of these massive bats have wing spans of five feet or more. Their large heads, in fact, sometimes resemble those of foxes or similar mammals. The creatures feed mostly on fruit and have excellent vision.

Another Case of a Common Name Gone Wrong

Common names given to organisms are sometimes confusing and misleading. Very often, they reflect some legend or superstitious belief of local people. An excellent example may be found in the common name given to some species in the genus *Lampropeltis.* These reptiles are still called "milk snakes" in some parts of the world, based on an old belief that they feed on cow's milk. Milk snakes, in fact, feed on rodents and other snakes. No documented case of them feeding on milk from a cow, or from any other animal, is on record.

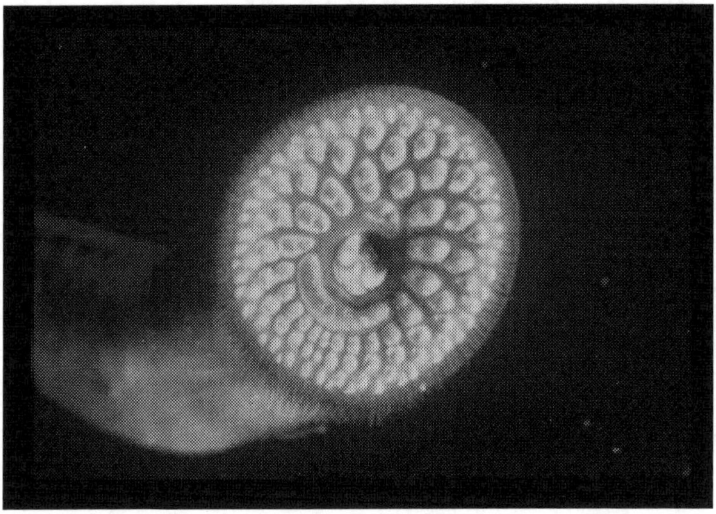

Detail of mouth of lamprey. Courtesy of USFWS.
Photographer not identified.

Fish without Jaws

Traditionally the biological class Agnatha (also called Cyclostomata) is regarded as the most primitive group of fishes. They lack jaws and have circular, sucking mouths. The two best known members of this class are the lamprey and hagfish. Lampreys are found in both salt—and freshwater habitats. Some species, in fact, migrate from salt water to fresh water for purposes of breeding. Lampreys are known to attach to host fish and feed on their blood. None of the few known species routinely exceeds two or three feet in length. Hagfish are also called "slime eels" by some people. They mostly act as scavengers, feeding on dead or dying organisms on the ocean floor. Adult lampreys rarely exceed three feet in length. Hagfish usually range between one and two feet in length. Larger ones have been reported however. Lampreys can secrete profuse amounts of slimy mucous when threatened.

Archerfish

At least five species of fishes, found from Asia to Australia, share the common name archerfish. They earned this name due their amazing ability to knock small prey items, like insects, from overhanging vegetation into the water. They achieve this by directing a thin, arrow-like stream of water from their mouths toward the prey. The jets of water are sometimes powered by enough force to travel more than ten feet from just below the surface of the water. Only the most skilled or lucky archer is able to hit a prey item from this distance, however. Archerfish hunt most effectively when food is located within a spitting distance of about six feet.

Flying Squirrels

More than twenty species or varieties of mammals that are collectively known as "flying squirrels" are found in the United States. They are also known from South America and in other regions. The name is a misnomer as these animals do not actually have wings and are not capable of true flight like bats and birds. What these organisms can do very well is glide. They possess hair-covered skin flaps which are spread to allow the animals to easily parachute from high to low tree branches and, occasionally, to the ground. The squirrels use their tails to help stabilize these short gliding episodes. They can scamper along the ground and among tree branches very effectively. Flying squirrels are almost always active only in late evening, night or early morning hours. One of the rarest flying squirrels is *Glaucomys sabrinus coloratus*, commonly known as the "Carolina Northern flying squirrel." This subspecies has been observed in only two states and five counties. Sightings of the organism are extremely uncommon in the mountains of Western North Carolina and Eastern Tennessee.

Swimming Unicorns

An unusual aquatic mammal known to biologists as *Monodon monoceros* has been given the vernacular names "sea unicorn" and narwhal. The unusual common name of sea unicorn comes from the fact that most males maintain a single tooth into adulthood that develops into an elongated tusk. It is usually a left-side, upper tooth that grows into the tusk. The tusks may sometimes reach nine or more feet in length. Females very rarely produce a tusk. Narwhals are found mostly in frigid Arctic waters. Adults may reach approximately 15 feet in length, with males being slightly longer than females. Another unusual feature of the narwhal is its habit of swimming and floating on its back for extended periods of time.

Plug in the Fish

Many species of electric fishes, both fresh water and marine, are known. The various species earned their common name due to the fact that they possess highly specialized nerve and/or muscle cells that generate unusually high levels of electric current. One species, in fact, may generate up to 50 volts of electricity. Fish utilize this electric current for a variety of purposes. Depending on the circumstances and species electric fish may use the electrical energy they generate to help them navigate through dark water, to stun their prey or to defend themselves. Some species seem to monitor disturbances in their electric field in order to help them detect a prey item or to evade a predator. Electric catfish, electric eels, electric rays and other species are known.

Squirrel Language

Squirrels are rodents that appear to have adapted especially well to habitats associated with people. Depending on how the common name is applied, more than 250 species are recognized. They are common in parks and yards, and in remote forests. The barking sounds of squirrels emit probably serve as communication. Also, biologists believe these mammals signal to one another with their tails. Evidence suggests that rapid movement of the tail serves to communicate threat or nervousness, while the act of holding the tail close to the back probably serves as a signal to other squirrels that no danger is present.

A Bait-Using Fish

An unusual fish, the angler fish, actually uses a lure and a line to help it capture its prey. Female anglers (in the genus *Linophryne*) have a tiny fish-like bit of tissue suspended from a thin thread-like appendage near their heads. The anglers move the lure about in order to attract the attention of small fishes upon which it preys. These fish swim close to the lure to investigate and are captured by the angler fish. Most angler fish are found at great depths, more than a mile in some cases, in the sea. Males are extremely tiny in comparison to the females.

Snow Monkeys

Snow Monkeys, *Macaca fuscata*, also known as "Japanese Macaques," are among the fastest learners of the order Primates. Episodes of food washing and bathing for long periods of time in hot springs were initially observed among very few individuals in isolated populations. With the passage of time, however, these behaviors appear to be spreading and are becoming much more commonly observed. Biologists suspect that novice macaques may imitate these behaviors after observing others in their group performing them.

Swimming Roaches?

Roaches do swim. However, common names can often confuse people trying to make sense of the natural world. The roaches in question are not the insects that are also known as cockroaches. Two species of rather unrelated fishes are commonly called roaches. *Rutilus rutilus* is a European species, similar to bream or bluegills. In North America, the term roach (when applied to a fish) often refers to *Hesperoleucus symmetricus*. This species is known from the west coast of the United States.

The Duck Billed Platypus

Native people of Australia have, of course, known of the platypus (*Ornithorhynchus anatinus*) for years. It is one of only two types of egg-laying mammals known. Other common names for these organisms include the "duck mole" and the "water mole." European scientists began to evaluate descriptions of this curious mammal in the late 1790s. A platypus skin was sent to the British Museum to document the organism's discovery by scientists. During this period of history, many hoaxes involving the natural world were being perpetrated. The museum curator immediately suspected that the platypus skin was just another such hoax. Believing that someone had glued or sewn a duck's bill onto the skin of a small mammal, he tried unsuccessfully to disjoin the bill from the skin with scissors. The curator's attempts to discredit the find failed. Today, scars from his work are still evident on the original specimen. Adults grow to about two feet in length and have a surprisingly soft, pliable bill. The male platypus actually produces potent venom that it injects by way of a spur-like protrusion. It is apparently used only for defense.

Lumpy Fish

The scientific name *Cyclopterus lumpus* has been given to a species of fish commonly called the lumpfish or the lumpsucker. These curious fish are found in the Atlantic Ocean and usually reach about nine inches to one and one-half feet in length as adults. Their skin colors range from dark green to purple or black. Numerous warty growths are common on the organism's skin. Both their common names and scientific name are derived from the fact that these fish use their ventral or belly-surface fins to form a sucking, disk-like attachment to rocks. The bond created is strong and not readily broken by outside mechanical forces. On occasion, the lumpsucker may attach to other larger fish. The animals are good swimmers. Males of the species are known for tending and guarding of the thousands of eggs produced during mating.

Where is the Cow's Home?

If the cows actually came home, where would we find them? The domesticated cow, *Bos taurus*, is familiar to almost everyone. It has been kept as a farm animal for so long that its ancestry has proven difficult to trace. The best evidence is that it was derived from *B. primigenius*, a now extinct ox-like animal that roamed wild in Asia and Europe prior to the 1600s. Today cattle are kept as livestock in many, many parts of the world. Well over 600 varieties or breeds are officially recognized.

Eating Like a Bird

Don't let this old saying fool you. Contrary to popular opinion, birds are big eaters. As a group, birds have a high rate of metabolic activity that requires tremendous amounts of food. In fact, birds spend almost all of their waking time engaged in activities related to eating or finding food. Some birds have very specific kinds of diets and may be adapted to eat only one or a few types of food. Other birds are more general in their feeding habits and may take several types of fare. The class Aves, which includes all birds, feeds on a tremendous variety of materials. Examples include green plants, insects, pollen, seeds, fruit, nuts, rotting flesh, fish, worms, rabbits, other birds, snails, snakes and tadpoles.

Beavers

In the United States and Canada, the mammal known to biologists as *Castor canadensis* is usually called the beaver. They are the largest known members of the order Rodentia (commonly called rodents) in North America. These organisms are widely associated with aquatic environments such as streams, ponds and lakes. They build dams and lodges with sticks, logs and mud. In doing so, they help to create and maintain wetlands. They may inadvertently cause problems for farmers and home owners with their work as well. Beavers feed on a wide variety of plant materials including underwater plant tubers, apples, leaves and tree bark. For feeding and building, a beaver's incisor teeth are the primary tool. They grow throughout the animal's lifespan and are usually kept at a length of about one inch. However, they may grow up to eight inches per year. The work they go through keeps the teeth worn down. The long, flattened tail assists the animal with swimming with and standing upright, on its hind legs. Beavers often stand upright. Contrary to popular belief, beavers do not carry mud with their tails. They will, however, move mud with their mouths and bodies toward the sound of running water. Some biologists have seen beavers throw mud on a tape player that was blasting a recording of the sound of running water. Beavers ordinarily mate for life. Their young often stay around to help with the next generation of offspring before moving away to establish their own territory and den.

American bison, often called buffalo. Courtesy USDA-ARS. Photograph by Jack Dykinga.

Buffalo

In North America, the common name "buffalo" is often applied to those herds of cow-like organisms that once roamed the great plains of the United States. The alternative common name, American bison, is more consistent with the organism's scientific name, *Bison bison*. Both sexes have horns. Massive herds of these organisms (numbering in the tens of millions) once extended from Canada down into the Southeastern United States, even into the Smoky Mountains. By about 1890, however, only about 500 of the animals remained on the planet. Many were killed in the 1800s by sportsmen or by individuals attempting to make life difficult for several Native American groups who utilized the buffalo as a dietary mainstay. Conservation efforts began in the early 1900s. Today the numbers of American Bison have increased. This legendary beast, sometimes weighing one and one-half tons and standing up to six feet tall, was for several years cast in likeness on five cent coins known to collectors today as "buffalo nickels." The old coins are mostly out of circulation. However, a similar pattern that features a buffalo was recently minted in the United States.

North America's Only Pouched Mammal

Its scientific name is *Didelphis virginiana* and it is the only known marsupial or pouched mammal in the United States. This organism is commonly called the opossum or possum. These very familiar mammals are about the size of a house cat. They range in color from white to gray to black and are generally active only at night. They have long, nimble tails that can wrap around a branch to help support the animal's weight. When threatened or injured, opossums enter an almost seizure-like state of paralysis that is mediated by the animal's nervous system. This is known as "playing possum" in many Appalachian communities. Possums have a large number of teeth and a very varied diet. They have been known to sometimes kill and eat many species of poison snakes. The opossums appear to be immune to the venom.

The Bower Bird

One of the most unusual sets of mating behaviors may be observed in the male satin bower bird, of the genus *Ptilonorhynchus*. These organisms spend a great deal of time during the mating season in the construction of basket-like structure called a bower. They carefully manipulate and bend small twigs and other materials into an arch. The purpose of the bower is to attract potential mates. Once the bower is built, the male bower birds decorate it with an amazing display of items. They will gather feathers, flowers and other brilliantly colored objects to place around and upon the bower. Blue appears to be an especially favorite color for decorations. In recent years, an amazing number of man-made items have been found at bower sites. Brightly colored hair clasps, rubber bands, buttons, pen caps and ribbons are just some of the many bizarre examples. The male bower birds appear to be in such an intense competition for mates that they will actually attack and destroy the bowers made by other males near their territory. Female satin bower birds appear to select their male partner based on the size and appeal of his bower. Males sing, jump about and display colorful bower decorations when a female approaches. Mating takes place within the bower. Males may mate with many females. Oddly enough, the bower is not the nesting site. Females leave the bower after mating and lay their eggs at another location. Males do not appear to assist in the nesting activities.

Porcupine Quills

Porcupines are mammals. They use genetically modified hairs, known as quills, as a defense mechanism. The hairs are stiff and tipped with a barb. Despite widespread stories to the contrary, porcupines do not fire their quills in self defense. They are unable to do so. What they can do is to strike toward the source of danger with their tails. If the barbs of their quills become lodged in the skin of a predator, they are left behind when the tail is drawn back.

Fish Breathing

All fishes have gills. Yet, it may be surprising to learn that some do not breathe exclusively by way of gills. Gills are feathery tissues, rich in blood vessels, which allow various organisms to exchange respiratory gases with the water or with very moist air. Some fishes have lungs. Since water cannot hold as much life-sustaining oxygen as air, it is a very common practice for these fish to periodically rise to the surface of the water to gulp air.

Bats

Those people who deplore bats may be surprised to find what interesting, intelligent creatures they truly are. Bats make up the largest order of mammals, Chiroptera, a term meaning "hand wings." One out of every four mammals is, in fact, a bat. Bats are the only mammal having the true ability to fly like birds. Some other mammals may be able to glide from trees. Bats are able to fly because they have a wing-like mechanism made of skin and modified bones of the hand and arm. Another surprise to most people is the fact that bats have well developed vision. Some species can see better than others. Like most other mammals, all bats bear their young alive. Females produce milk for their nourishment. In some species, the baby bats cling to the mother as she flies in search of food. The dietary habits of bats are extremely varied. Many species do us a great service by feeding on insects. Some eat nectar or pollen and help to pollinate various types of plants in the process. Some bats are fruit eaters. Others feed on fish or smaller animals that they capture in much the same fashion as a hawk, eagle or other bird of prey. Only three species of bats, all commonly called the "vampire bat," are known to feed on blood.

Look Owl Below

Speotyto cunicularia may prove to be one of the strangest birds known to biologists. It is commonly known as the "burrowing owl." The owl earned this title due to its habit of excavating a tunnel, sometimes up to ten feet long, to protect its underground nesting site. It is the only known bird that nests so deeply underground. The burrowing owl may sometimes remodel existing tunnels and dens made by other species of animals. The species is found mostly in the Western United States with some populations reaching South America. They average about nine inches in height.

Seahorses

Seahorses are fish with modified bodies and a grasping tail-like appendage. At least fifty species in the genus *Hippocampus* are known but none exceeds more than a few inches in length. In all known species, female seahorses transfer their ova into a brood pouch in the male seahorse's body. The male then fertilizes the ova and incubates the developing babies until they are born alive. During the incubation period, the developing young derive some oxygen and a few nutrients from their father.

The Sperm Whale

Although they are not the biggest known whales, sperm whales are giants. Males of the species *Physeter macrocephalus*, sometimes reach lengths of more than 65 feet. These immense toothed whales have an unusually large head. A cavity in the head contains abundant quantities of a wax-like material called spermaceti. Early whalers thought that it was stored sperm. While the purpose of spermaceti is poorly understood today, it is believed to either have something to do with the whales' ability to maintain balance or with its ability to navigate through water. Another product from sperm whales that was prized by industry in early times was called ambergris because of its golden-brown color. Ambergris has a strong, sweet odor and has been used in the manufacture of perfumes.

Fish That Carry Nurseries in Their Mouths

The practice of holding fertilized ova or juvenile offspring inside the mouths of parents is known as mouth brooding. Several species of organisms do this, including many kinds of fishes. Quite a few species within the family Cichlidae, known as cichlids, are especially noted for this activity. Cichlids are bony fish that are strictly found in fresh water, primarily in the southern hemisphere. Many are kept in aquaria as pets. *Tilapia esculentas* parents hold fertilized eggs in their mouths. *Bujurquina vittata* go a step further and actually keep newly hatched offspring within their mouths during times of danger. In some species of cichlids, both parents mouth brood. At the sign of a predator, the tiny young fish may race inside the mouth of their much larger parent for protection.

CHAPTER FIVE

GENETICS & EVOLUTION

A Living Fossil Fish

For many years scientists knew the coelacanth, a fish, only from fossil remains. Scientists assumed that the entire genus had suffered extinction more than 100 million years ago. A specimen was caught off the coast of Africa in the 1930s, much to the surprise of western scientists. The curator of a museum in the region documented the catch. Since then, several other living coelacanths have since been found. Today most biologists recognize two species of the ancient fish, both classified in the genus *Latimeria*. Some specimens of coelacanth have been measured at nearly six feet in length.

How Many Chromosomes?

Chromosomes are discrete, condensed bodies of DNA which contain genes. Humans generally have 46 chromosomes in each of their body cells. The number is halved to 23 in the reproductive cells which are called ova and sperm. Chromosome numbers vary widely among living organisms. Interestingly enough the number of chromosomes an organism has in its body cells has no relationship to its complexity or intelligence. Following is a list of the chromosome number for a variety of organism: Tobacco 48, some frog species 26, carrot 18, horse 66, starfish 30, domestic chicken 78, soybean 40, some crayfish species 200, the crayfish *Astacus trowbridgei*, 376; some species of pear 68, the fruit fly *Drosophilia melanogaster* 8, domesticated dog 78, domesticated cat 38, redwood tree 66; the microscopic, unicellular protist *Amoeba proteus* 600.

Larvae . . . Not Just For Insects

Many, many animals begin their life cycle as larvae. Larvae (singular is larva) almost always look and behave very differently from the adult organisms they will become. They are sexually immature and tend to live in different habitats and feed on different types of food than do the adults of the species. Strangely enough, the term larva is derived from a Latin word meaning ghost. In addition to some insects, there are several other organisms which possess larval stages during their life cycles. They include starfish, frogs (called tadpoles), ticks, mussels and many others.

Pitcher plants growing in the wild. Courtesy of USFWS.
Photographer not identified.

Pitcher Plants and Cobra Lilies: Evolved to Capture Insects

Pitcher plants and cobra lilies are insectivorous plants with leaves, modified to form a tube, containing water and digestive enzymes. Insects are usually lured to the plant by the odor of the flowers. They enter the tube and become disoriented by light (in some species) or are prevented from escaping due to downward pointing hairs and/or a slick interior surface of the pitcher. The trapped animals may take a few days to several months to be partially digested by the plant. The largest of these plants only hold about a quart of fluid. The biggest documented prey items include frogs, mice and rabbits. Like all insectivorous plants, pitcher plants and cobra lilies make their own food through photosynthesis. They catch insects and other small animals to supplement the mineral-poor habitats in which they usually grow.

Other Carnivorous Plants

In addition to pitcher plants and cobra lilies, other types of carnivorous plants have been discovered. These include the Venus fly trap which catches small insects and other tiny animals in trigger-released traps made from leaves; as well as the butterwort and sundew plants which capture prey with sticky secretions. The bladderwort is a floating plant found in ponds and streams in certain parts of the world. Tiny balloon-like bladders open in response to touch to capture the prey. As previously noted, all carnivorous plants are known to carry out photosynthesis. They make their own food but rely on insects and other prey items to provide supplemental minerals, mostly nitrogen-based compounds.

Ancient Megafauna

The term megafauna literally means "large animals." Examples of modern day megafauna include elephants, giraffes, whales and large fishes. By studying fossils, biologists have been able to infer a lot of things about large animals of the past. Countless genera and species have been named. In some cases, complete fossilized skeletons have been found. In other cases only a few fragments of bone or teeth have been unearthed. Several well preserved specimens of mammoths have been recovered from glacial ice. Only a few of the many interesting mammals that belong to the past are mentioned here. The cave bear, *Ursus spelaeus*, was much larger than our modern bear species. However, scientists believe that the cave bear ate mostly plant material. Its remains have been found only in Europe. The oldest fossils are dated to about one-half million years old. The cave bear was probably extinct by about 10,000 to 20,000 years ago. *Acinonyx pardinensis* was the giant cheetah of Europe and Asia. Other species of this fearsome predator lived in North America. One of the most famous cats of the ancient megafauna is the saber tooth cat. These powerful hunters lived in North and South America between about 1,500,000 and 10,000 years ago. They have been classified in the genus *Smilodon*. Several species of giant sloth have been identified from the fossil records. As a group, they lived between 30,000,000 and 8,000 years ago. Some species may have been up to 20 feet long. *Megalonyx jeffersoni* left fossils in the Western United States. A better known relative, *Megatherium americanum*, inhabited South America. The largest known land mammal was *Indricotherium transouralicum*. It has gone by a number of common names since its discovery in the early 1900's. One fairly descriptive moniker is "giant hornless rhinoceros." The oldest fossils of *I. transouralicum* date to about 25 to 30 million years ago. It was probably extinct by about 15 million years ago. Based on fossilized reconstructions of the animal, some individuals probably reached heights of 16 to 18 feet, from the foot to the shoulder. They had long necks and probably ate leaves from tall trees and other plants.

Hybrids

In one use of the term, a hybrid organism is the product of a mixing of genetic information from two different organisms. If you are reading this book you are, in fact, a hybrid that resulted from combining genetic information from your mother and your father. The word hybrid is also widely used to describe the offspring from two different species of organisms. Hybridization of this type occurs very rarely in nature due to various sorts of ecological, behavioral and mechanical factors that keep species separated in terms of reproduction. When hybrid organisms are born they are almost always sterile, if they survive at all. Farmers and scientists have worked to produce a number of hybrid organisms for curiosity and for economic reasons. Some examples follow. A plant often known as the cabbish resulted from artificial crossing of a radish and a cabbage. A hybrid derived from a female horse and a male donkey is called a mule. Interestingly enough when a male horse and a female donkey are mated the offspring differs slightly. It is called a hinny. Ligers or tiglons result from breeding a lion and a tiger. The beefaloe is a product of a cow and a bison, also commonly known as a buffalo.

Complex Bee Genetics

Common honeybees, *Apis mellifera*, are native to Europe and are only one of a number of species of social insects that have complicated genetics and social structures within their colonies. Their method of sex determination is known as haplodiploidy. In this system, all females are derived from fertilized ova produced by a single queen. All males are produced from unfertilized ova and have only half the number of chromosomes as their sisters. The sisters end up sharing about 75 % of their genetic material, 25 % more than organisms that do not utilize haplodiploidy. Potential new queens in a colony are fed a nutritious substance called royal jelly. This extra nutrition gives the females the ability to become fertile, capable of reproduction. Usually only one new queen will survive. All worker bees are sterile, having not been fed royal jelly. Worker bees find food, clean the hive, feed the larvae, expand and defend the hive, and use their wings to maintain air flow throughout the hive. Only workers and queens are capable of stinging. Male bees are called drones. They are driven away from their home colonies after they reach sexual maturity. Once they mate with a queen from another colony they will be stung to death. The males are capable of mating only once because of the fact that they actually inject their reproductive organs into the tip of the queen bee's abdomen.

The Unassuming Lancelet

They look like small worms at first glance. Closer inspection may sway the observer into thinking that she is looking at a small fish. The lancelet, however, is neither worm nor fish. These tiny, brownish creatures seldom exceed two inches in length. Biologists are fascinated by lancelets because they display features that link them to both invertebrates, like insects and worms, and to vertebrates such as fish and reptiles. Lancelets are also known by the common name amphioxus. They belong to the phylum Chordata; the same phylum into which humans classify themselves. Unlike humans and other vertebrates, however, the lancelet does not have a true back bone. It does retain a supporting notochord throughout its life. Lancelets often burrow into the gravel and mud where they filter food from the water. They derive oxygen from the water by way of gills and have a well developed digestive system.

Canis familiaris

His scientific name literally means "the common dog" or "the servant dog, familiar to everyone." Our pet dogs (taken as a whole) represent a wonderful lesson in genetics, evolution and selective breeding. The origin of *C. familiaris* remains unclear. Humans have been associated with the domestic dog since ancient times. Many biologists believe that *C. familiaris* was originally the product of selective breeding and hybridization of various species and varieties of wild dogs of the genus *Canis*. The wolf, *Canis lupus*, is thought to have played a particularly important role. An extinct wolf-like organism known as *Tomarctus*, as well as various species of foxes, may also account for much of the modern dog's heritage. Throughout the course of time, countless breeds or specialized varieties of the common dog have been developed through selective breeding or by genetic chance. All members technically now belong to the same species and have 78 chromosomes. However, the argument can be made that extreme sizes among the various breeds of domestic dog would prohibit mating between some breeds. Some dog breeds have an interesting history. A few are listed below.

A Few Interesting Dog Breeds

The Labrador retriever was likely developed in Newfoundland and was prized for its ability to trail and recover game animals during hunting. Beagles continue to be prized for tracking. They were known in historical periods as early as Medieval England. The breed known as the "German shepherd" probably originated in Europe. It has served man faithfully as a guide dog, law enforcement dog and as a guard for farm animals. Basset hounds, developed in France in the 1600s, were also prized for tracking and hunting. The same is true for the great dane and the Jack Russell terrier. The former is thought to have originated in Germany. The latter was named for John Russell, a gentleman from England, who promoted the breed. Dalmations came from Dalmatia, in Croatia. They were so sporty that they often traveled with horse-drawn carriages and coaches as ornamentation. They are sometimes called the "coach dog" for this reason, not because of their black and white spots. They have historically been honored as mascots in firehouses as well. Fox terriers have an English ancestry. They were prized for their ability to drive foxes from their dens. The greyhound may date back as early as ancient Egypt. It has been used in racing and hunting for centuries. Many historians believe that the Chihuahua was used centuries ago as a sacrificial animal in Mexico. Common poodles likely originated in Germany. They were valued for their ability to hunt truffles and retrieve birds. Even in these early times, poodles were regarded as having grand showmanship and were often trained to perform various tricks for the amusement of their masters. Doberman pinschers are known to have originated in Germany in the last years of the nineteenth century. They were named for the dog breeder, Louis Dobermann, and were prized as police dogs and watch dogs. The Cantonese butcher's dog is today known as the chow or "chow chow." In ancient China, these animals were valued as guard dogs and for pulling sleds. They were also used as watch dogs and for food and clothing. Monks in Switzerland kept St. Bernards trained to assist with rescues. The origin of the Shih Tzu is unclear. It was likely derived from ancestors in Tibet and China as early as the 600s. The animals were pampered as royal pets in China. Border collies are known from the 1800s in Great Britain. They remain noted for their ability to manage heard animals on the farm. Boxers were valued in Germany in the 1800s for their fighting prowess and for guarding the home. Yorkshire terriers appear to have been selectively bred in Great Britain to hunt and kill rats in cotton mills. Dogs known as komondors are valued for their ability to protect farm animals, especially sheep. They are native to Hungary. The breed called the dachshund was known as the "badger dog" in early Germany. It was prized for its ability to remove badgers from their deep, narrow holes. Regardless of their genetic heritage, all dog breeds are valued today as loyal pets by countless millions of people.

Fossils of Many Types

Fossils are the preserved bodily remains of ancient organisms or other types of preserved evidence of their existence. Fossils are generally considered to be rare finds. Their formation was likely happenstance. It should be noted that some organisms probably have never left behind any sorts of fossilized remains due to their lack of durable body tissues. Also, in some parts of the world, fossils may never be formed due to lack of various geological and environmental factors that appear to be necessary. Most of us are familiar with fossilized bones or petrified tree trunks. Scientists believe that these types of fossils were formed slowly as minerals began to enter the organism's remains and assume its shape. In other cases, organisms may have become suddenly buried in mud or other sediments to form fossils. These types of situations may have also preserved footprints of various sorts of animals. Small insects and the like were perhaps trapped in tree resins which hardened into a glass-like display case to preserve the organisms. Finally, some whole animal carcasses, such as wooly mammoths, may be found deep-frozen in ice. Other animal remains may be found in tar pits.

The Genus *Homo*

Humans have classified themselves into the genus *Homo*, meaning man. In fact, we called ourselves "wise man" when we adopted the scientific name *Homo sapiens.* Mainstream anthropologists believe that other members of the genus lived on Earth in times past. Evidence for these ancient men comes from fossils, tools and other sources. The naming of ancient fossil remains is a controversial task. Some evolutionary biologists contend that several species may have existed in past times. *H. halibus* (handy man) was named for his ability to use many types of stone tools. The common name "upright man" is reflected in the scientific name *H. erectus.* Other named members of the genus include *H. rudolfensis* and *H. ergaster*. Some scientists insist that the famous Neanderthals belong to the same genus as modern humans. The name *H. neanderthalensis* has been suggested to denote these ancients. A few evolutionary scientists recognize two further species, *H. antecessor* and *H. heidelbergensis* that are traditionally grouped within the Neanderthal assemblage. Many other scientists place the Neanderthals in a separate genus. Mainstream biologists believe that modern humans are the only living members of the genus *Homo* on the planet today. Fossils of the ancient members of the genus are dated back as far as two million years ago and as recently as about 30,000 years ago. Some of these ancient species probably coexisted at various points in time.

Humboldt penguins. Courtesy of USFWS.
Photograph by Dr. P. Dee Boersma, University of Maryland.

The Evolution of Two Strange Birds: Warm Weather Penguins

Most people tend to associate penguins with frigid Antarctica. That is the first image many people call to mind when these mostly flightless, black and white birds are mentioned in conversation. It is of note, however, that several species of these interesting birds exist throughout the world. Some are found in Australia, some in Africa and even in Chile and Peru on the South American Continent. The South American species is sometimes called the Humboldt penguin, *Spheniscus humboldti*. It may reach heights of a bit more than 25 inches. A species that uses its wings for swimming in the ocean is known from the Galapagos Islands. The Galapagos penguin, *S. mendiculus* is the tiniest penguin identified by biologists. It seldom reaches more than about 15 to 20 inches in height and is a fast swimmer, reaching speeds bursts of up to 25 miles per hour. Of all species of penguins, it is the most northerly known and lives at the equator. Some biologists believe that the ancestors of both *Spheniscus* species were carried north, by the Humboldt oceanic currents, from Antarctica in the distant past.

Some Reproductive Terminology

Various methods of sexual reproduction exist in the animal kingdom. Organisms that lay eggs which develop outside the female are known as oviparous egg layers. The eggs may be fertilized internally in some species and externally in others. Examples include birds, grasshoppers and many fish. In organisms that are ovoviviparous, females retain the eggs internally and bear live young. As with oviparous organisms, the developing embryo derives nutrition from the yolk sac of the egg. The dogfish shark and many species of snakes carry out this type of reproduction. Viviparous organisms are similar in their development except for the fact that the female provides some nutrition to the developing babies, supplemental to the yolk sac. The common guppy, a popular aquarium fish, reproduces in this fashion. "Amniote egg" is a general term for any fertilized egg or ovum that is enclosed in a leathery or hard shell. This protective shell allows the species not to be as dependent on water for reproduction, as would be the case with frogs and fish which lack amniote eggs.

Two Ancient Species Saved From Extinction

Evolutionary biologists generally believe that more than 98% of the species that have ever lived on the planet are now extinct. In modern times humans have certainly contributed to, or directly caused, the extinction of many species. However, evidence suggests that people were probably responsible for saving two tree species from extinction. At one time the ginkgo tree (*Gingko biloba),* the dawn redwood (*Metasequoia glyptostroboides*) and their close relatives were known to science only from the fossil records dating at least as far back as the time of the dinosaurs. Both trees were later found in very remote areas of China and were proclaimed to be living fossils. The word ginkgo loosely translates from a Chinese phrase meaning "the silver apricots." Evidence suggests that both of the trees probably were extinct in the wild but were kept in cultivation in monastery gardens and similar situations. Both species are the only known survivors of their genera. They are widely planted today and are even available through nursery mail order catalogues.

Single Ancient Landmass

Many historical geologists believe that giant landmasses called supercontinents have come and gone throughout the history of the Earth. Most recently, between about 200 million and 300 million years ago, most all of the land on Earth appears to have been compacted into a giant super continent known as Pangaea. Evidence suggests that Pangaea began to slowly break apart shortly thereafter. Giant plates of the Earth's crust may have moved to form two smaller landmasses known as Laurasia, to the north, and Gondwana to the south. As the drift of the land masses continued, the continents and islands that we know today probably assumed their current positions on the globe. This theory of continental drift is important to the theory of evolution because groups of animals and plants may have been isolated on particular land masses for millions of years while undergoing their own courses of evolution. Most evolutionary scientists favor this scenario to explain unusual and characteristic distribution patterns of various organisms on Earth today.

He and She, Boy and Girl

Organisms that reproduce sexually generally fall into two categories. In the case of many plants and simple animals, one organism may produce both male gametes (sperm cells) and female gametes (ova). Life forms that do this are referred to as hermaphroditic or monoecious. Those that produce only one type of gamete or sex cell per organism are called dioecious. In the case of dioecious organisms, you can never go wrong by referring to the partners as male and female. What about other species or group-specific names for the male and female? Some interesting examples are listed below. Males are listed first, females second. Most birds are referred to as cocks and hens. For chickens, the male's name may be changed to rooster. Male turkeys are often called toms or gobblers. Ducks go by drake and duck; falcons by terzel and falcon; hawks by tiercel and hen. Geese are referred to as gander and goose; swans as cobs and pens. Dogs and their relatives are tagged as dogs and bitches. In the case of foxes, the handles often are changed to reynard and vixen.

More He and She, Boy and Girl

Most people would associate the terms bull and cow exclusively with cattle. Other organisms that share these designations include the yak, camel, hippopotamus, elk, rhinoceros, seal, walrus, moose, dolphin, porpoise, whale, elephant, alligator, crocodile, buffalo, bison and many others. Giraffe males are called bulls but females are called does. Some biologists refer to male sharks as bulls. Male caribou are sometimes called harts and male oxen bullocks. In a variation, the terms buck and doe are often assigned to gerbils, hamsters, rabbits, squirrels, antelopes, rats and mice and deer. Male deer may be called stags. Boar and sow terminology belongs to pigs, hogs, hedgehogs, raccoons, prairie dogs, bears, panda bears, badgers and sometimes guinea pigs. Horses and their relatives (such as zebras) may be called stallions and mares. The term dam is a more obscure term for the female that is still sometimes used. Sheep may be called bucks and ewes. Male goats sometimes are known as bucks but may often be called billies. Females are nannies or does. The same is true of kangaroos but the terms boomer and jack are frequently used for males; flyer, roo and jill for females. Jack and jill may be used in the case of the wombat, wallaby, opossum and other organisms. For donkeys the terms may remain the same or be changed to jack and jenny or jackass and jennet. In some uses of jack and jill, the male's name may be changed to hob. Such is the case with ferrets and weasels. Tom and queen are generally used only for house cats. For larger female cats the names change to such things as lioness and leopardess. Male social insects such as bees, yellow jackets and wasps are invariably called drones. Females that are capable of reproduction are known as queens. Those that remain sterile are called workers. Bacteria are, of course, not capable of true sexual reproduction. They are asexual. Some species can, however, exchange segments of genetic information. In cases such as this, microbiologists often refer to the bacterial cell that donates or releases the genes as positive (+) and the cell that receives the information as negative (-).

A Giant Ape Man?

Giant ape-like creatures are frequent stars of fantasy and fiction. Their possible existence in the past has tantalized scientists and laypersons for years. Between the 1930s and the 1950s, Western scientists examined a few fossilized teeth and jaw remnants that suggested a link to such a creature. These studies, and others, alluded to the profile of an ape that stood up to ten feet tall and weighed more than 600 pounds. The organism was named *Gigantopithecus blacki* and was thought to have inhabited ancient Asia in the distant past. Most mainstream scientists believe that the organism was extinct by about one million years ago. However, some cryptozoologists have proposed that the creature may still be living in small populations in Asia and elsewhere in the world. *G. blacki* could, in the opinion of some, represent the elusive creatures known as Bigfoot and Yeti.

A Male Midwife

One of the strangest methods of parental care known is that of the midwife toad, *Alytes obstetricans.* These small gray-brown-green frogs are known from various parts of the world. Their mating practices are not that remarkable. It is what follows mating that has earned the frog its common name. Females produce 20 to 80 eggs that are fertilized by the male. The eggs are firmly cemented to the father's rear legs. He drags the large egg cluster with him for several months. When the eggs are about to hatch the father frog enters water, usually a pond, where the eggs hatch and the tadpoles swim away. It is also of note that both male and female midwife toads secrete toxins from their skins for purposes of defense from predators.

Ancient Reptile Order

Biologists have classified several extinct reptiles into the order Rhynchocephalia. Only two species, in a single genus, are known to exist in the world today. *Sphenodon punctatus* and *S. guntheri* are known as sphenodons or tuataras. Fossilized remains of both species, and others, have been found in New Zealand. Living members of both species are today known only from small nearby islands. These lizard-like creatures may live up to 100 years but they require between 10 to 20 years in order to reach sexual maturity. Sphenodons live in burrows and are usually active only at night. They feed on insects and sometimes eat small birds or bird eggs. Some may reach about two feet in length.

Change of Identity

In many simple organisms hermaphroditism is a way of life. In this method of reproduction one individual produces both male gametes called spermatozoa and female gametes called ova. Organisms that reproduce in this fashion are also called monoecious. Self fertilization is rare in such organisms. A few organisms are actually able to change their sex due to various genetic and environmental influences. Protandry involves a male organism changing to a female. Some species of seabass, and the anemonefish, are known to carry out this practice under certain conditions. Protogyny is female changing to male. Two species of fishes, wrasses and sea goldies, are examples. Some flowering plants often have their male and female flower parts developing at different times, thereby effectively changing their sex. It should be noted that the terms protogyny and protandry may also refer to cases in which one sex within a population matures sooner, completes migration more quickly or is somehow otherwise separated from the other in time or space.

Thunderbirds: Myth or Reality?

Almost all native tribes in North America have rich oral traditions, religious practices and histories involving gigantic birds that made sounds like thunder during flight and that preyed upon humans and other large animals. These creatures are collectively known as thunderbirds. In fact many ancient civilizations made cave paintings and other likenesses of these creatures. Historians and cultural anthropologists tend to refer to these creatures in strictly mythological ways. Some paleontologists, however, have made discoveries that lend a possible element of reality to these stories. Very few people believe that remnant populations of giant flying reptiles could have sparked the stories in ancient times. Others believe that a group of giant birds, collectively known as teratorns, could have been a reality. A likely contender has been reconstructed from fossil remains found in California and named *Argentavis magnificens.* Some biologists estimate that this giant, vulture-like bird had a wing span that regularly exceeded 15 feet and may have reached up to 25 feet. It is estimated to have stood about five feet tall. Some believe that it was capable of true flight, or at least could glide efficiently for great distances. The mainstream view is that *A. magnificens* and other teratorns became extinct about 10,000 years ago. Stories about giant birds have been recalled and recorded since well before colonial times in North

America. A string of scattered but poorly documented accounts have survived into the 1800s and even into modern times. Some cryptozoologists believe that populations of giant birds, fully capable of attacking and eating humans, may remain in North America and in a few other locales on the planet.

A Different Method of Sperm Transfer

In a few species of organisms, males prepare their sperm cells in enclosed capsules called spermatophores. Some spermatophores are very simple packages but others have especially elaborate shapes. These probably help to eliminate the possibility of hybridization with other species in the wild. In some species, the male partner may physically transfer the packet of sperm into the reproductive tract of the female. This method is rather common in insects and other small invertebrates. Squids follow this procedure as well. In vertebrates like newts and salamanders males often leave the spermatophore stuck to a leaf, rock or other substrate for the female to pick up. In this case, she manipulates the packet of sperm into her own reproductive track opening. Sperm are received and stored in specialized receptacles known as spermatathecae. In most cases of sperm transfer by spermatophores, the sperm-containing capsule begins to swell and contract once inside the female. This will help ensure release of sperm cells.

Sexual Selection

Charles Darwin was the first biologist to formally propose the idea of sexual selection. In short, sexual selection is believed to influence evolutionary patterns due to traits that make a potential mate more likely to be successful in reproducing. As a general rule, male animals and birds usually display the effects of such selection. Intrasexual selection may involve a competition between two or more males to determine which one will mate with female(s). On the other hand, intersexual selection is ordinarily determined by female choice among potential male mates. Examples of traits that may have been driven by sexual selection include the brilliant coloration of some birds, massive antlers of mammals, unusual mating calls and oversized appendages.

Artistic rendering of *Polyodon spathula, the* American Paddlefish. Courtey of USFWS. Image created by Timothy Knepp.

Odd Fish

The fish family Polyodontidae is of note to evolutionary biologists for several reasons. Members of the family are most commonly referred to as paddlefish. They are classified with other bony fish but their skeletons are mostly made of cartilage. Superficially, they resemble sharks but are not closely related to them. These massive fish tend to be gray in color and have elongated, flattened snouts. Modified tissues on the underside of these paddle-like snouts serve as taste buds. The fish lack true teeth but capture food through a number of specialized structures near the back of the mouth called gill rakers. Paddlefish swim with their mouths open in order to trap food in the gill rakers. Most biologists believe that fishes in the family Polyodontidae were common in ancient times. Only two genera and species are known to survive today; one in China and the other in the United States. *Psephurus gladius* is the Chinese species and lives in the Yangtze River. Some specimens have been recorded with lengths in excess of six feet. The American species is *Polyodon spathula.* It may exceed eight feet in length and 200 pounds in mass. This species lives in the Mississippi river system and has a large flattened snout. These are some of the most unusual freshwater fish known to man.

Life from Nonlife

Spontaneous generation is an ancient belief that living organisms can originate from non-living matter or from the breakdown of other types of living things. In older times, laypersons and scientific thinkers alike tried to make sense of their world based on their own observations. They constructed what today seem like fanciful accounts on the origin of living things. For example, it was widely taught that a horse hair placed in water would turn into a worm. Some people believed that frogs originated from mud and that mosquito larvae fell to the Earth within raindrops. Some people claimed that a mixture of cloth, water and wheat kept in a container would turn into mice within several days.

Hunt For the Ivory Billed Woodpecker

Campephilus principalis is the scientific name for the ivory billed woodpecker. It is the largest woodpecker of modern times in North America. These beautiful black and white birds may reach up to 20 feet in length. Males have a brilliant red crest on their heads. That of the female is black. Both sexes feature an all-white, elongated bill. Until recently the ivory billed woodpecker was regarded as being either a highly endangered, or a recently extinct, species. A few eager people often mistake the pileated woodpecker, *Dryocopus pileatus*, for the ivory billed. Pileated woodpeckers are slightly smaller but do superficially resemble the elusive ivory billed species. Sightings of animals in the wild can be terribly hard to document and confirm. Such is the case with the ivory billed woodpecker. The species was probably never very common. Biologists believe that they require large expanses of old growth forest, preferably near swamps and streams, in order to feed and reproduce. Recently dead trees, infested with insects, are their primary food source. The practices of logging and irresponsible land development are probably most responsible for the demise of these beautiful birds. By the 1940s the ivory bill was extremely rare in the United States. Some sightings in the 1980s and 1990s were reported. A few years ago, a possible series of encounters of remnant populations of the bird in Arkansas were discussed. Most biologists think that if the bird does survive elsewhere it is most likely to be found in a few tiny, scattered populations in Louisiana and Cuba. Some doubt that. At any rate, a few hopeful biologists and amateur birdwatchers frequent the known final ranges hoping to get a glimpse.

Genetics and Evolution in Action

In populations of African and Indian elephants (genus *Loxodonta* and *Elephas maximus*, respectively) there is a genetic variation in which organisms fail to develop tusks at maturity. Elephants utilize their tusks in many ways including display, food gathering and defense. This accounts for the fact that tuskless elephants were once rather rare. In some populations only about ten percent of tuskless adults could be censused. Fast forward to the booming ivory trade that developed in relatively recent times. Many poachers kill adult elephants, remove their tusks and leave the carcasses behind. Since this human-induced pressure has been introduced into elephant populations, one could argue that being born without tusks is an advantage. Elephants without tusks may be at a disadvantage in terms of food gathering and defense but they are at no risk for poaching by ivory hunters. It makes sense, then, that tuskless elephants would be more likely to survive and reproduce than those with tusks. The number of tuskless elephants in many populations has increased dramatically to more than 25%.

Complex Kangaroo Reproduction

Marsupials, those mammals that protect their developing young in specialized pouches, are a noteworthy group. Within this group, the kangaroos have even more peculiar reproductive habits. Three large species within the genus *Macropus*, all found in Australia, are usually called kangaroos. Other species within the genus are sometimes called wallabies. *M. giganteus* is the Eastern gray kangaroo. *M. fuliginosus* is the Western Gray and *M. rufus* is the red kangaroo, the largest of all. Newly born kangaroos tend to be a little more advanced, in terms of their development, than are the other marsupials. The babies are still rather tiny, however. Once they emerge, they crawl into the female's pouch and attach themselves to a nipple in order to nurse. One of the most remarkable things about the female is that she can produce three or four different types of milk to match her offspring's stage of development and nutritional needs. The youngest are fed milk that is rich in sugar and low in lipids. As the offspring age and advance, the sugar content of the milk declines while lipids increase. All of this becomes even more amazing when it is noted that a kangaroo mother may simultaneously be nursing a new born and an older juvenile that only stays in the pouch for short periods of time. A kangaroo baby is, by the way, called a joey. In some species of kangaroos development of the newly fertilized ovum may proceed only for a very short period of time and then become arrested or frozen in place. The tissue mass will resume development when a nursing joey dies or begins to become less dependent on its mother for nutrition. The average joey spends up to 8 months within its mother's pouch and becomes more independent as it ages. It has been said many times that the name of the kangaroo was derived from an interesting exchange between native Australians and Western explorers, many years ago. As the story goes, an explorer pointed to a group of kangaroos and asked some Aboriginal inhabits what the animals were called. Their response, sounding something like kangaroo, was probably intended to indicate that they did not understand the question they had been asked.

Geologic Time Scale

Most scientists believe that the Earth is about four and one-half billion to five billion years old. Rock formations and fossil beds have been given names on a geological time scale. The scale is divided, according to some versions, into four eons. Eons are divided into various eras. The eras may be subdivided into periods to further clarify and categorize trends in the geological and fossil records. Periods are comprised of epochs and epochs of ages. The oldest eon containing recognizable fossils is the Archean, derived from a Greek term meaning "ancient." This layer dates back to about three and one half billion years ago. The Proterozoic (meaning "early animal") eon began almost 550 million years ago. These two spans, along with the older rock layers that do not contain fossils are often collectively referred to as the "Pre-Cambrian supereon." Many scientists think that the end of the Pre-Cambrian was marked by a mass extinction of algae. The next major division of note is called the Paleozoic era. This term means "old life." It began about 540 million years ago with the Cambrian period. The Ordovician period began about 485 million years ago. Once again, scientists suspect that a mass extinction event occurred on the planet to eradicate the majority of the genera of marine invertebrates. The Silurian period began 445 million years ago, the Devonian 420 million, the Carboniferous 360 million and the Permian period about 300 million years ago. The close of the Permian period marks, according to many scientists, the most massive extinction event in the entire history of the Earth. The Mesozoic (middle animal) era followed the Paleozoic. It has been called the age of dinosaurs and began about 250 million years ago with the Triassic Period. Next came the Jurassic period about 200 million years ago; and the Cretaceous period 145 million years ago. At the end of the Cretaceous period, evidence suggests that a major catastrophic event (or a series of such events) occurred on the earth. This marks the last time that dinosaurs appear in the fossil record. The Cenozoic (new animal) era came next and began about 65 million years ago with the Paleogene period. The Neogene period came next. Finally, the Quaternary period started about two and one half million years ago.

The Famous Fruit Fly

Probably the best known and most widely studied species of fruit fly is *Drosophila melanogaster*. These tiny flies are only about three millimeters long. Since the early 1900's, they have been the subject of exhaustive research on genetics and development. They continue to be a favorite research organism due to their small size and well-studied genetics. Their entire life cycle, from fertilized egg or zygote to adulthood, may take as little as 10 days to complete. *D. melanogaster* have eight chromosomes and somewhere in the neighborhood of 14,000 genes. Like other flies, they are noted for the giant (or polytene) chromosome clusters that appear during their larval stage. These exceptionally large concentrations of DNA are valued by geneticists doing research on how genes are expressed. *D. melanogaster*, and other species within the genus, produces some of the largest sperm cells known in the animal kingdom. The fruit flies are commercially sold for research purposes to everyone from elementary school teachers to professional biological researchers. Mutations occur in a wide variety of easily observable traits. For example both wingless (apterous) and vestigial-winged flies are kept in pure culture. Other wing mutants include lanceolate or elongated wings, scalloped, curved and curly wings. The eyes of the wild-type fruit fly are red. Mutations involving the eyes incorporate the color variations of white, cherry, wine, apricot and sepia (brown/red). Eyeless mutants also occur. A gene called the period gene influences the flies' sleep/wake cycle and emergence from the pupal case. It may be mutated in a number of ways so that *D. melanogaster* behaves as if it were on a cycle either shorter or longer than the normal 24 hour cycle. The shape of the flies' body bristles is also under genetic control. Wild-type bristles resemble human eye lashes. Mutations include split bristles, shaven bristles that appear to have been cut, and spineless bristles that tend to flatten and curl. Numerous other genetic varieties of *D. melanogaster* are known.

Bioluminescence

A wide variety of organisms display the interesting property of bioluminescence, the ability to generate their own light. Many species of fungi, fish, insects, protists, jellyfish and other organisms perform this amazing feat. The glowing light emitted by these creatures is a by-product of their metabolic activity. In the presence of oxygen, a chemical called luciferin is activated by one of a category of enzymes, collectively called luciferase, to produce light energy. The name luciferin was probably derived from ancient words meaning light-bearing. The term lucifer is historically associated with the planet Venus (the morning star) and, of course, with Satan from Christianity. While the purpose of the bioluminescence in some species is unknown, it is well documented in others. Depending on the species the light may be used as a signaling or communication device, a lure for food or a means to confuse predators.

How Long Until Ready? Plants

Some organisms are able to breed almost immediately after they are born. For most organisms that reproduce sexually, however, there is a waiting period during which growth and development of the reproductive organs must occur. Some plants reach sexually maturity within only a few days or weeks. Others take longer. Environmental conditions such as sunlight, space, soil quality and other variables may impact the necessary amount of time. Pecan trees may take from five to 20 years to bear nuts. Avocado trees are usually reproducing by the age of six years. The creosote plant and the ceanothus, a desert shrub, may take up to 15 years to reach maturity. Saguaro cacti and gingko trees sometimes delay sexual maturity up to 30 years. The grand champion of the plant kingdom seems to be the sequoia tree. Some individuals probably require up to 200 years in order for them to become reproductively viable.

How Long Until Ready? Animals

Amounts of time required to reach sexual maturity vary widely within the animal kingdom as well. In some cases, males and females of a species may require differing periods of time for sexual development. Horseshoe crabs may require up to 10 years in order for them to reach sexual maturity. Fish like the Atlantic bluefin tuna are usually ready in eight years. Sturgeons may be ready that quickly as well, but sometimes take up to 25 years. The brown shark reaches maturity about 16 years after birth. Among the reptiles, development also varies widely. Snapping turtles usually reach maturity within 18 to 20 years. Galapagos tortoises typically take 20 but green sea turtles require 25 to 30 years. Sphenodons range from 10 to 20 years to attain the ability to reproduce. Timber rattlesnakes may take up to nine years, and ball pythons a much shorter two to four years. Some birds are ready to reproduce in as little as one year. Penguins may take three to eight years, depending on the species and other variables. Camels are sometimes able to reproduce in just three years. Other mammals may take longer. For example gray whales may take between five to 11 years and bottle nose dolphins between six and 11. This wide range is accounted for, in part, by the fact that females mature much earlier than males. Asian elephants are often able to reproduce within about 10 years of their birth.

Preserved specimen of the passenger pigeon.
Courtesy of USFWS. Photograph by Luther Goldman.

An Unlikely Extinction

Who would even imagine that a species that numbers into the billions of individual organisms could become extinct? Unfortunately, it has happened. The saga of the passenger pigeon,

Ectopistes migratorius, is one of the most dramatic and sad ecological stories of all. The passenger pigeon is said, by many, to have been the most common bird on the Earth. Some biologists estimate that there were more passenger pigeons in the wild than all other species of North American birds combined. Several credible written reports and descriptions of the organism remain, along with a few preserved specimens. The birds were excellent flyers and could reach speeds of up to 70 miles per hour. They often traveled in gigantic flocks that could include hundreds of thousands of organisms. Some reports indicate that the sky grew dark as these huge assemblages passed overhead. Accounts of some flocks exceeding 250 miles in length and one mile in width are deemed accurate. Large trees are said to have been broken down by the weight of thousands of passenger pigeons. The demise of the passenger pigeon was clearly due to over hunting. Some people valued the bird as food. In many cases, however, they were shot entirely for sport and left dead. In some well organized hunts, tens of thousands of passenger pigeons were killed within a small area in one day. Some hunters even captured a live passenger pigeon, sewed its eyes shut, and tied it down to serve as a powerful decoy to attract other birds. It is said that this practice was the origin of the phrase "stool pigeon." The last documented sighting of a tiny flock of passengers came in 1910. The last known bird died in captivity in 1914.

Kitty's Chromosomes

The house cat, known to biologists as *Felis domesticus*, (or more properly *Felis catus*) has been associated with humans for centuries. The exact origin of the house cat has been lost in time. Many believe that modern day house cats may have their genetic origins in a wild African cat, *Felis silvestris*. Ancient Egyptians kept the cat in domestication. It even had a place in their religious observances. House cats have been in North America since early colonial times. The various breeds and varieties of house cats seem to be endless. Cat breeders are at an advantage if they understand the genetics behind such traits as hair color and hair length. While the cat's genome has yet to be completely mastered by humans, a few things are known. One of the most recent traits to have been involved in selective breeding is caused by a dominant gene that produces curled external ears. The color of cats is a complex matter. Many genes interact to produce a cat's coloration. There is a gene that codes for a uniform hair color. When this gene is expressed, the cat will not display Siamese or Burmese coat patterns. There is another gene, one form of which signals for the development of tabby coat patterning. Still another gene will code for yellow banding on individual hairs. It is called the agouti gene by geneticists. Black fur may be achieved by way of the presence of a dominant gene. Its recessive alleles may produce chocolate and cinnamon coat colors. All white-colored fur in house cats may occur in two or three ways. First the cat may suffer from albinism, the absence of pigment in the hair, skin and eyes. A more common method involves the interaction of genes for white spotting with other genes. Two copies of the gene may produce all white fur. It is of note that, contrary to widely held beliefs, all white cats are not deaf. Deafness in white cats is correlated with the presence of blue eyes. The genetics behind this correlation are not well understood and only a small percentage of these cats are actually without hearing. Another interesting characteristic of cats, under the control of a gene, is reaction to catnip. Many cats go into a behavioral frenzy in the presence of the plant. They are believed to have a dominant allele controlling the reaction. Those cats that do not obviously react to catnip odors are believed to have recessive alleles for this gene.

Large Flightless Birds

The ostrich, *Struthio camelus,* is the largest known living bird. It is found mostly in Africa. Individuals can exceed eight feet in height and weigh more than 300 pounds. Moa is a common name given to a group of extinct, herbivorous flightless birds from New Zealand. They ranged in size from species as small as a turkey to others that exceeded nine feet in height. *Dinornis robustus* was probably the largest species. All species of moa were most likely extinct by about 1600. However, there are unsubstantiated reports of these birds that came from the 1850s. The dodo, *Didus ineptus*, was found living on the island of Mauritius in the Indian Ocean. It is believed that this bird's range was restricted to that island alone. The dodo was slightly larger than a wild turkey and is believed to have reached extinction in 1681. Many historians believe that the bird was valued as food and hunted by humans to extinction. However, more recent theories suggest that non-native species were mostly to blame. It is believed that dogs, pigs, rats and other animals brought to Mauritius disturbed the dodo's nests and fed on their eggs and young. It should be noted that some scientists think that there were actually two or three other species that lived at the time of the dodo, but on other islands. They may have belonged to the same genus.

More Large Flightless Birds

The great auk, *Pinguinus impennis,* also known as the garefowl; grew to about two and one-half feet in length on average. These birds once had a wide-ranging habitat that extended from Canada into Florida. The great auk was hunted to extinction for its plumage and oil by the mid 1840s. They were good swimmers but could not fly. The tooth diver bird (*Hesperornis regalis*) was probably an excellent swimmer. Only fossil records of this bird survive, as it became extinct by about 65 million years ago. This bird was probably about six feet long and likely hunted fish in the ancient seas. The elephant bird, *Aepyornis maximus*, is believed by many to have been the largest bird ever in terms of body mass. It reached heights up to 10 feet and its eggs measured more than one foot in length. This flightless bird was found in Madagascar and was probably extinct by 1650. Other living flightless birds include the endangered members of the genus *Casuarius*, commonly called the cassowary. These birds are found in New Guinea and Australia. They may exceed six feet in height. The South American rhea, *Rhea americanus*; and the lesser rhea, *R. pennata* are also flightless. The larger bird may reach five feet in height. Members of the genus *Apteryx* are relatively small, about the size of a chicken. They are found in New Zealand and are commonly known as kiwis.

How Closely Related Are You?

For organisms that reproduce sexually, like humans, geneticists often speak of coefficients of relatedness. This term essentially reflects the percentage of genes that two individuals are likely to share. Barring unusual cell division processes (which do sometimes happen) offspring should share about 50% of their genetic information with each other and with each parent. Aunts and uncles share about 25% of their genome with nieces and nephews. First cousins share an average of 12.5%.

Mastodons and Mammoths

Is there a difference between these two ancient groups of elephant-like mammals? In fact there are several. Mastodons belong to the family Mastodontidae or Mammutidae. The family name, in part, comes from the complex tooth surface patterns found on fossilized remains. Many species of the family had tusks on both the upper and lower jaws. Fossilized remains of some tusks are more than 15 feet in length. Mastodons were probably browsers, animals that eat plant material from trees and shrubs. They lived between about 10,000 and four million years ago. Mammoths belonged to the family Elephantidae, the same as modern elephants. They were probably grazers, animals that eat plant material from the ground. They were larger on average than mastodons, and lived between about 4,000 and 120,000 years ago. Some scientists insist that mammoths were alive as long as two million years ago, however. The tusks of some species were close in size to those of mastodons. Not all mammoths were wooly mammoths. Some probably lacked the long, thick hair for which mammoths are often noted. Most evolutionary biologists believe that neither mastodons nor mammoths were direct ancestors of any of our modern elephant species. They may have shared common ancestry, however.

Strange Method of Sex Determination

In a worm of the genus *Bonellia,* there is an unusual method of sex determination that has been documented. The worm inhabits oceanic waters. Juvenile worms are essentially indistinguishable from one another, in terms of sex. If a juvenile enters the reproductive tract of a mature female it will become a male and continue to live there, in the uterus. If it remains outside of a mature female, it will become a female.

A North American Parakeet

In the recent past North America had one known species of parrot, *Conuropsis carolinensis*, commonly called the "Carolina parakeet." By all accounts, the bird was beautiful. Including its elongated tail feathers, it averaged a length of 13 inches. Brilliant combinations of green, orange, yellow and blue added to the bird's splendor. The Carolina parakeet was most common in the Southeastern United States. It ranged as far west as Texas, and possibly beyond. Humans drove the bird into extinction for two primary reasons. The first was the bird's showy feathers. They were used in hats and for other decorative purposes. Also, the animal was regarded as a pest to the growing farming communities within the bird's range. The birds fed on a wide variety of fruit, seeds and grains. Untold numbers of Carolina parakeets were shot or poisoned until the species vanished. The last bird in captivity died in 1918. A few wild organisms may have lived into the 1920's in Florida.

Famous Finches

Darwin's finches, also called the "Galapagos finches," are a group of 15 species of birds the size of small sparrows. As a group, they were discovered by Charles Darwin during his visit to the Galapagos Islands and neighboring land masses in the 1800s. However, it appears that these birds had been mentioned in the writings of earlier explorers in the late 1790s. Darwin collected a few specimens from the islands but wrote very little about them. Many modern evolutionary biologists consider this group of birds to be a classic case study in adaptive radiation. It is thought that all island species were descended from a single species on the South American mainland. The ancestors may have been blown to the islands during a storm or delivered by some similar occurrence. Since the islands were rich in unexploited food sources, as some believe, the species may have evolved to occupy various niches. The species differ mostly in their beak form and in the foods they utilize. Some species feed on leaves; others use their beaks to dig into trees in the same way as woodpeckers. Still another species utilizes a cactus spine to fish insect larvae from their refuges. Some species feed on insects, fruit and many other types of foods. There is considerable disagreement about the classification of Darwin's finches. Traditionally they are placed into one of four genera, though some taxonomists recognize five. *Geospiza* includes the ground finches. *Camarhynchus* includes the tree finches and *Certhidea* the warblers. The genus *Pinaroloxias* includes a single species found on a nearby island, not in the Galapagos island group.

Genetics and the Notion of Subspecies

Sometimes, within the group of organisms that biologists call a species, there are populations of individuals that are different enough in terms of their genetic makeup that they are sometimes called races, varieties or subspecies. They probably could reproduce with all members of the species but still retain some uniqueness that is noteworthy enough to earn a subspecies distinction. A good lesson in subspecies genetics may be found in the species known as *Panthera leo*, commonly called lions. According to most biologists, this species had as many as seven to nine distinct varieties with the recent past. Most people tend to associate lions with Africa. That is where most of them are found. Fossil evidence for an extinct American subspecies, *P. leo atrox*, has been uncovered. A living subspecies known as *P. leo persica* is found today in small numbers in India. It has been called the "Indian lion" and the "Asiatic lion." This organism was probably very wide spread on the Asian continent at one point in time. Its range likely extended into Eastern Europe as well. Today it is known to exist in the wild in only one Indian forest. Including those in captivity, only a few hundred survive. The Asiatic lion tends to be slightly smaller than its African cousins. The males have a smaller and less dense hair mane that leaves their ears exposed. In terms of behavior, the Asian subspecies tends to live in smaller social groups than do those in Africa. By using sophisticated genetic analysis, some biologists have claimed that the African and Asian lions are more closely related (in terms of their genes) than are the three main races of human beings on the planet. It is estimated by some that the Asian and African populations of *P. leo* may have become geographically isolated less than 100,000 years ago.

When Reptiles Flew

Flight is not listed among the characteristics of reptiles in modern biology books. However, an ancient group of flying reptiles of the order Pterosauria has been described from fossil remains. As a group these creatures are more commonly known as pterosaurs. They are believed to have been capable of flight due to the presence of a pair of wings that were composed of thin skin stretched over modified forelimbs. The pterosaurs lived during the age of the dinosaurs and likely became extinct around 65 million years ago. A few poorly documented accounts of modern flying reptiles exist; none are taken very seriously by mainstream biologists. From the fossil record, several genera of flying reptiles have been named. Among the smallest reptiles thought to be capable of flight was *Sordes*, estimated to have reached a length of about one and one half feet. *Eudimorphodon* was a slightly larger relative that reached just less than two and a half feet in length. Probably the best known pterosaur was *Pteranodon*. They are believed to have reached lengths of about 20 to 25 feet and had wings spanning twenty or more feet in width. Many paleontologists believe that *Pteranodon* was eclipsed in size by a flying reptile that almost reached double its size. This creature, *Quetzalcoatlus*, is known only from a few fragments of fossilized remains. Some researchers think that it could have reached lengths of up to 40 feet and had a wing span that may have been as much as 45 or more feet.

Giving Birth Out of the Mouth

Rhinoderma darwinii was named for one of the scientists who first described it, Charles R. Darwin. We know this unusual frog today as "Darwin's frog." It is green and black in color and usually about one inch long. The frog's head terminates in a sharply pointed snout that it uses to help it breathe while it floats on its back in streams and ponds. The organism's unusual shape and swimming habits should be enough to keep biologists interested, however these frogs have another trick up their sleeves (or should we say in their mouths?). After mating with his partner, the male Darwin's frog guards the fertilized eggs for a period of about two weeks. As he begins to see significant signs of development within the eggs, he will take them into his mouth. The father does not swallow the eggs but sends them to the vocal sac at his throat. The eggs hatch into tadpoles that are nourished by the yolk sacs from the eggs. The tadpoles continue their development into tiny frogs while inside their father's vocal sac. When they reach about half the length of the parents, they are allowed to emerge from the father's mouth. This phase of development may take between one and two months. Imagine Charles Darwin's surprise when he saw a male frog giving birth to babies from its mouth.

CHAPTER SIX

ECOLOGY

A Beetle from Japan

Most people in the United States call them "Japanese beetles." Biologists know them as *Popillia japonica*. These brilliantly colored green pests feed on a wide variety of plant materials including the leaves, flowers and fruits of many plant species. They can attack a plant with such savagery that defoliation and even death are often the result. Japanese beetles are known for their ability to multiply very quickly, and in great numbers. After mating, the females burrow several inches into the soil to lay their eggs. The larvae will, of course, give rise to more beetles. These insects have been in the United States for about one hundred years. Most biologists and historians agree that a few individuals were probably accidentally introduced here on imported plant specimens. They gained a quick foothold. Japanese beetles remain pests in the United States, with few natural enemies, even today.

Food from Animals

It is certainly no secret that humans utilize many, many animals as food. Mostly, we eat muscle tissue which we call meat. However, many other types of foods are derived from animals. Bees and other insects of their order make honey. Nectar is ingested from flowers and changed to honey in the bees' esophageal sacs. Honey is mostly a mixture of water, sugar and oil. A single worker bee routinely makes less than one tenth of one teaspoon of honey during her entire life time. Cheese, butter and yogurt are made from the milk of cows, sheep and other mammals. An enzyme called chymosin is often used in the making of cheese because it helps milk to curdle. It may be derived from certain stomach contents (called rennet) of calves or other young mammals. A food known as chitterlings is enjoyed in some cultures. It is prepared from the small intestines of hogs. Usually it is deep-fried in grease. The internal linings of mammalian stomachs, particularly cows and goats, are known as tripe and are used as food. Foie gras or pate is considered to be a delicacy by many people. It is made of goose liver. Head cheese, also known as souse or souse meat, is a favorite in some parts of Appalachia. It is a cooked, seasoned mixture of various hog parts including (but not limited to) the head, ears, feet and snout. Liver mush is made from cooked, seasoned hog's liver. Finally, a snack food known as pork rinds is made from deep fried hog skin.

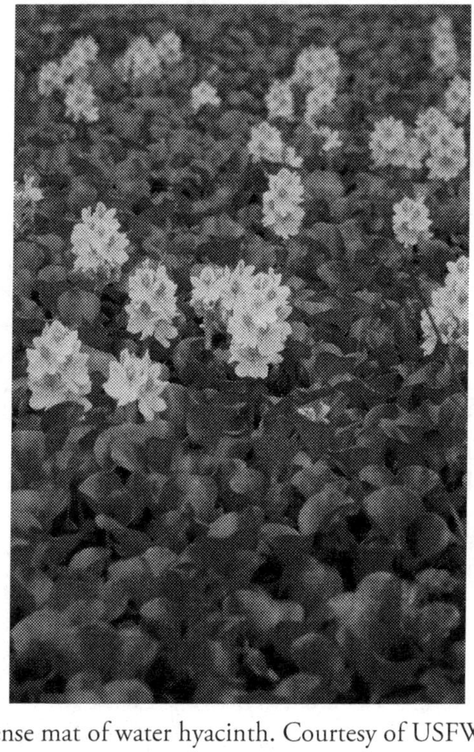

Dense mat of water hyacinth. Courtesy of USFWS.
Photograph by V. Ramey.

A Beautiful South American Menace

Many freshwater streams, ponds and lakes in South America sport one of the most beautiful and interesting plants known to science. These small plants, *Eichhornia crassipes*, float on the surface of the water and display showy purple blooms. The common name is "water hyacinth." In the 1860s specimens of *E. crassipes* were brought into the United States as ornamentals. Many people began to grow them in ponds and even in container-based water gardens. Before much longer, the introduced plants started to become a serious nuisance. The water hyacinth can reproduce both asexually and sexually. It does so very rapidly. Very few animals in North America will feed on the plant. Even today, the plant remains a source of problems. Many waterways have become clogged and overgrown. In Florida, manatees (one of the few mammals that eat water hyacinth) have been moved from stream to stream on many occasions in an effort to control the growth of the menacing but beautiful plant.

Symbiotic Relationships

Many organisms live in such close associations that an entire category of terms has been invented to classify and describe their relationships. In general any close association between two different species is known as symbiosis or as a symbiotic relationship. There are three basic types: mutualism, parasitism and commensalism. Mutualistic relationships occur between two organisms that both derive benefit from the association. Bacteria live in the intestines of many species of leeches, and help them to digest blood. Parasitic relationships occur between two organisms, the host and the parasite. The parasite derives some benefit and the host is harmed. Aphids, tiny insects, feed on the sap of various plants. Commensal relationships occur as one member of the association is benefited and the other is unaffected. Tiny arthropods called barnacles often live on the skin of whales. They apparently do no harm to the whales but are provided with a secure home site.

Living Long

The maximum length of an animal's life span is often very difficult to document. It is a generally accepted rule of thumb that most animals tend to live longer in captivity than in the wild. Reports of longevity among captive animals are often exaggerated. Very few cases of humans surviving past the age of 120 are accepted as legitimate. Interestingly enough, scientists estimate that ancient Neanderthal people rarely survived up to the age of 20 years. The following list includes maximum life spans of various animals, wild and captive, that are generally accepted as accurate by mainstream biologists. An unusual marine organism known as "the giant tubeworm" is believed to live for up to 200 years and possibly beyond. Some species of giant tortoise have lived up to 180 years. Turkey buzzards may rarely survive for 118 years, and some species of swans and clams live for 100 years or more. Scientists estimate that blue whales and some sea anemones may live for around 90 years, some species of eels and freshwater oysters for 80. Domesticated cockatoos and Indian elephants have documented maximum life spans of 70 years. Ostriches, turtles, alligators, catfish and horses may survive for 60 to 65 years. Geese, orangutans, chimpanzees, eels and some lobsters have maximum life expectancies ranging from 50 to 55 years. A few instances in which domestic cattle and pet goldfish have lived more than 40 years have been recorded. The odd pairing of house cats and canaries shares a maximum life span record of 34 years. A polar bear was documented with a life span of 33 years. A rare case of a domestic dog surviving a little more than 29 years has been recorded. Mountain lions and kangaroos may sometimes live more than 20 to 25 years. Beavers, domesticated sheep and rabbits rarely, if ever, survive past their teens. The queens of some ant colonies may live for up to 15 years. Chickens may survive up to about 14 years. Hummingbirds, with their extraordinary metabolic rates, are considered very old if they live for eight years. Finally, some insects may live for only a few days or hours.

Taxic and Tropic Responses

Living organisms respond to their environments in a variety of ways. Tropic responses are generally considered to be growth responses; they are most often associated with plants. Taxic responses are more associated with animals and tend to involve faster movement than simple growth. An organism may display either a positive or a negative response. In the case of a positive response, the growth or movement is toward the stimulus. Negative responses are away from the stimulus. Some specialized types of responses have been given particular names. A phototaxic or phototropic response is a response to light. The shoots and leaves of green plants are positively phototropic while the roots are negatively phototropic. Earthworms try to avoid light by crawling away and are therefore negatively phototaxic. Gravitropism, sometimes called geotropism, is a growth response to gravity. Plant roots generally grow toward the pull of gravity; stems and leaves grow away. Thigmotropism involves the sense of touch. A plant may grow toward a material that it could climb on by way of tendrils, for example. Some plants drop or wilt their leaves in response to touch. This action likely serves to protect the plant from being eaten. Carnivorous plants may monitor stimulations of touch, and react to them in order to capture food. Chemotaxic and chemotropic responses are common among many organisms as they seek out or avoid chemicals such as acids, oxygen, pheromones or any of a wide variety of other chemicals. Responses concerning water are known as hydrotaxic or hydrotropic responses.

Caution Lights

Many species of animals, particularly amphibians such as frogs and salamanders; and insects such as yellow jackets and lady beetles, display prominent warning colors. Throughout the animal kingdom it appears that very bright colors such as red, orange and yellow are universally approached with caution. Most of the organisms that display this warning coloration secrete toxins that can make predators sick or leave them dead. A few are only mimics. They display the colors but lack the toxins.

Competitions in the Wild

Competition for limited resources in the environment takes place daily in the world. Even plants may compete for space and sunlight. When the members of a species of organisms in one particular area are competing for a resource such as food, shelter or nesting sites, ecologists call that intraspecific competition. If the contest is between two different species for the same resources, interspecific competition is said to be occurring. Biologists do not believe that two species can hold exactly the same niche at the same time within a community. Most evidence suggests that when different species within a community of organisms utilize the same resources, they often do so in slightly modified ways. This can help to reduce the negative effects of interspecific competition.

Playing Dead

The term thanatosis has been given to the act of faking death. Many organisms pretend to be dead as a defense mechanism. The idea is that a potential predator may be repelled by a dead prey item, thereby saving the organism's life. North American opossums are probably the most famous animals that play dead. When threatened or startled, opossums become limp due to a seizure-like response from their nervous systems. Hog snakes and a few other species of snakes open their mouths and roll over to expose their bellies in a pose that may bluff a predator. When forcefully rolled, tiger sharks also become apparently lifeless for several minutes at a time. Many species of insects and spiders may also become stalled in death-like positions when threatened.

A House of Straw (and Dirt)

The thickly growing perennial grasses of the Western and Mid-Western United States have a strongly developed root system that holds soil in place very well. During the westward homestead movement of the 1800s, farmers found this sod to be extraordinarily resistant to plowing. The ground beneath was very fertile, however. Despite this menace toward establishing viable farms, pioneers found many uses for the sod. They would cut brick-sized portions of the grass and soil and use these to make their houses, barns and other outbuildings. In fact, it was not unusual for roofs to be sealed with these sod blocks. Because trees were so scarce, and lumber out of reach due to isolation, many of these "soddies" began to dot the landscape. A few of these original dwellings still survive as a testament to the tenacity of the western pioneers. Most native citizens of that era used materials such as animal hides or mud to build their homes.

I Saw a Bunch of Them

Many animals are known to live, travel and/or feed in groups. Some groups are very informally organized while others have well defined standards of membership. Examples of names assigned to some groups of animals follow. In general, a group of birds is called a flock or a company. More specific terms include a bevy of swans (or a wedge if they are in flight), an aerie of eagles, a murder of crows, a rafter of turkeys, a muster of peacocks, a dissimulation of wrens, a band of blue jays, a charm of hummingbirds, a nye of pheasants, a flight of swallows, a gaggle or skein of geese, a covey of quails, a dole of doves, a watch of nightingales, a herd of cranes, a flush or brace of ducks, a rookery of penguins, a parliament of owls, a congress of ravens and a descent of woodpeckers. Finally, a group of young birds in the nest is generally called a clutch. Fish are usually referred to as a school or shoal when traveling in groups. The term herd is often applied to gatherings of cattle, antelope, seals, walruses, sheep, elephants, deer, zebras, horses, giraffes and similar animals. Dogs, coyotes and wolves travel in packs; while hyenas move in clans. A group of foxes is sometimes called a skulk or an earth. Many types of insects such as wasps, bees and termites live in colonies but travel in swarms. The term grist is sometimes used to describe a group of bees. Other insects that do not have truly social organization may travel in groups called swarms or clouds. Examples include locusts, gnats, flies and moths.

More Animals by the Bunches

Monkeys are organized into a troop, gorillas into a shrewdness or band. Oysters, clams, starfish and similar types of organisms may exist in beds or colonies. Cheetahs form coalitions, lions form prides, housecats sometimes run in clutters; tigers organize to form streaks while leopards assemble in leaps or prowls. Pods may be made up of whales, dolphins or crocodiles. Prairie dogs live in towns. Rhinos travel and feed in crashes and snakes may den together in a pit or bed. There are drays of squirrels, labours of moles, mischiefs of mice, skulks of foxes, warrens of rabbits, droves of pigs, smacks of Portuguese men of war; flutters, smacks or smuks of jellyfish; sleuths of bears, knots of frogs, surfeits of skunks, bales of turtles and rafts of otters. Further, businesses of ferrets, hoards of gnats, knots of toads, stenches of skunks and gams of whales may be seen.

Eating Habits

Any organism (such as an oak tree) that can make its own food is called autotrophic. Those that must eat other organisms, or their remains, are called heterotrophic. Humans are heterotrophic. Other more specific terms exist to further classify an organism's eating habits. Carnivores eat animals, herbivores eat plants, omnivores eat a combination of plant and animal foods, insectivores feed specifically on insects. Larger animals, like vultures, that tend to feed on dead animals are often called scavengers. The terms saprovore and decomposer are mostly reserved for bacteria, fungi and other microbes that feed on decaying remains of other organisms. In terms of ecological relationships, plants (because they make their own food) are often called producers. An animal that eats these plants is called a primary consumer. A secondary consumer feeds on a primary consumer and a tertiary consumer feeds on a secondary consumer. Therefore, food made by the plants cycles through the ecosystem.

A Cleaning Service

Several species of large beetles in the family Scarabaeidae go by common names such as "dung beetle" or "tumble bug" due to their unusual habit of moving about balls of animal feces. Some species of this group may move a ball of fecal material for several yards toward a nest site. Females lay their eggs upon the fecal mass and the developing larvae feed on it. Feces may contain a lot of undigested, or partially digested, foods that will nourish the developing larvae. The few organisms that actually feed on solid wastes of other organisms are collectively known as coprophages.

The Dust Bowl

Years ago several states in the mid-western United States were given a nickname, "the dust bowl." These regions became plagued by extremely severe wind storms that carried tons of soil during a long period of years beginning in the 1930s. These storms were responsible for destroying farmland, buildings, equipment and livestock. Humans suffered a number of ailments during the storms, including difficulty breathing. The loss of income and homes was immeasurable. The causes of these severe dust storms are poorly understood. Scientists believe that a combination of drought, high winds and poor soil conditions (brought about by over cultivation) worked together to begin these destructive and unusual weather patterns.

View of Earth from space. Courtesy of National Aeronautics and Space Administration (NASA). Photographer not identified.

Down to Earth Trivia

Earth is the third planet from the sun. It has an estimated diameter of slightly less than 8000 miles and a circumference of nearly 25000 miles. The total surface are of the planet is about 196,935,000 square miles. Around 70% of the surface is covered by water. The highest point of elevation is Mount Everest at 29,028 feet above sea level. The Dead Sea exists at 1,302 feet below sea level. Some scientists estimate that the Earth is about four and one-half billion years old.

Do Other Animals Sleep?

Sleep appears to be very important to humans and most other mammals. The importance of sleep remains unclear to biologists but likely serves to allow an organism to carry out any number of cellular processes that lead to health and restoration. All mammals appear to have a sleep pattern that is close to that of humans. Birds and reptiles also appear to sleep regularly as well. For lesser organisms, however, the question of sleep is not so easily addressed. Some studies reveal that a state analogous to (but not exactly like) human sleep occurs in many insects, amphibians and fish. Other studies are inconclusive. With mammals, sleep can sometimes be correlated with lifestyle. As a general rule, organisms that are vulnerable to predation tend to sleep less often and for shorter intervals of time. Those that have few natural enemies appear to be able to enjoy much extended sleep times. Of course, exceptions to these generalizations do exist. Following is a list of average sleep times per 24 hour day for some organisms. Chimpanzees average about nine and one half hours while baboons and dogs sleep about 10 to ten and one half hours. Rabbits sleep about 11 hours per day. Cats are champion sleepers. The common house cat may sleep as many as 16 hours per day while the cheetah sleeps an average of 12. Lions average 13.5 hours and tigers 15.5 hours. Cows, horses and goats rarely sleep more than four hours. Giraffes sleep even less, with a two hour average per day. Some species of bats sleep up to 20 hours per day. Pythons keep pace with them, sleeping between 15 and 20 hours per day, on average. The North American opossum may sleep about 18 hours per day. Some species of dolphins slumber about seven hours per day. The duckbilled platypus of Australia averages 14 hours per day, the house mouse 12. Parrots and domestic chickens usually clock around nine hours per day. The Komodo dragon sleeps about 15 hours per day. Frogs do appear to snooze regularly during warm weather. There is a great range of seven to 18 hours per day, depending on the species. A common aquarium fish known as the guppy seems to sleep an average of seven hours per day. A species of fruit fly routinely undergoes a period of sleep-like activity that lasts about seven hours per day.

Flies That Play With Fire

In the Southern United States, there are two genera of fireflies that are involved in a remarkable predator-prey relationship.

Fireflies are not flies at all, but are actually beetles. Several species of fireflies in the genus *Photinus* breed at about the same time. The males flash their light emissions in patterns that are specific to their species. Females of that species may flash to capture the male's attention for potential mating. By using slightly differing flashing patterns, males of the various species of *Photinus* may find mates within their own species. Another genus of fireflies, *Photurus*, often occupies the same habitat as *Photinus.* Incredibly, female *Photurus* can mimic the flashes of *Photinus.* This allows her to get the male's attention and draw him toward her. *Photurus* females prey upon *Photinus* males looking for a mate.

Fresh Water and Marine Perils

Organisms that live in fresh water and those in marine or salt water environments face very different perils and pressures. This is due to the fact that water has a tendency to move into cells and tissues that contain a higher salt content than the water itself. This movement of water is known as osmosis. The tissues of fish that live in fresh water would explode if the water entering their bodies was not expelled. These fish get rid of the excess water by urinating. They almost never swallow water. On the other hand, fish in marine environments would quickly dehydrate if they did not replenish their water stores. These fish swallow water and tend to produce only small amounts of urine with a low content of water.

Where Parasites Roam

In general terms a parasite exploits another organism (called the host) for various things such as food, reproduction or shelter. A number of parasites may actually kill their hosts while others cause them minimal harm. Some parasites spend their entire lives in association with their hosts. Other types encounter the host only briefly during their life cycle. If a parasitic organism reproduces within or upon their host organism, the host is known as the primary host. If they are not associated with a particular host for purposes of reproduction, that host is called an intermediate host. About 25% of all known animals are parasites during all or part of their life cycle.

Night Shift and Day Shift

Due to various pressures from their environments, organisms tend to be active at different times of the day. The term nocturnal has been used to describe animals that feed, mate, hide from predators, and carry out other life activities mostly at night. Those that carry out their life activities during the day have been called diurnal. The term crepuscular has been used to describe organisms that have a propensity toward action during twilight hours, dawn and dusk.

Body Temperatures

We often hear the terms "warm blooded" and "cold blooded" applied to various species of animals. These terms can be very misleading. The terms ectothermic and endothermic are much more representative of the situations. Ectothermic animals include insects, fish, snakes and many other organisms. The body temperature of these organisms is largely influenced by the temperature of their environment. They are only cold blooded on cold days. If the outside temperature becomes too hot, these organisms are in danger of overheating if they do not seek shelter or otherwise cool themselves. Ectothermic organisms include birds and mammals. They are able to maintain a relatively constant internal body temperature that is not so heavily influenced by external factors. These organisms have a high metabolic rate that generates much body heat. In extreme heat or cold, their internal mechanisms for maintaining normal body temperature may be overwhelmed.

Habitat, Range and Niche

These three terms are often used when discussing the habits and distributions of particular animals. They are frequently confused. The term range refers to the geographical region in which a particular species may ordinarily be found. The habitat is more specific and refers to the sort of place where an organism lives. Examples of habitats may include a pond, grass, treetop or soil. Think of the habitat as the organism's physical address.

The ecological niche of an organism refers to the functions that organism carries out in its habitat. Niches may include preying on particular species of animal, gathering nectar from flowers or eating nuts from a certain tree. The niche of an organism can be thought of as that organism's job.

A Crabby Boxer

One of the most unusual cases of symbiosis involves a small marine crustacean known as the "boxer crab." These crabs appear much like any other. They have a strange method of protection and food getting, however. Boxer crabs superficially appear to have flower-like bouquets on the end of their front claws. These "flowers" are actually tiny sea anemones. Sea anemones are coelenterates that have stinging tentacles used to capture food. Oddly enough the boxer crabs place anemones on their own pincers, much like a boxer would wear boxing gloves. The crabs wave the anemone about and then nibble at the food that has been captured. Even though the anemones are small, their presence probably helps to protect the crabs. It is likely that the anemone benefit from this bizarre relationship as well. They are probably given greater opportunities to capture food by way of the association. A few species of boxer crabs have been described. *Lybia tessellata* is known from the Philippines and *L. edmondsoni* from Hawaii.

Echolocation in Bats

It is true that some species of bats use a form of echolocation, similar to SONAR, to navigate the night skies and find prey. Bats that feed on night-flying insects usually do this. When the bats are in flight they often send out very short, high pitched vocalizations that may reach beyond the range of human hearing. Bats are able to alter the rate and pitch of these sounds, depending on circumstances. When a sound wave from a vocalization strikes an object and echoes back, bats are able to judge their distance from that object as well as what the object may look like. Large objects may represent walls, trees or other potentially dangerous items. Smaller moving objects may represent flying insect prey.

Human Population Growth

About 1000 years ago, scientists estimate, there were less than one half billion people living on the planet. Today that number is estimated to be more than seven billion. Humans have mastered a number of things in their environment that account for this increase. For example, they have discovered or developed medicines that prolong the lifespan. They have also found methods to preserve foods and have done other things to improve their health and safety.

Is that a Horn or an Antler?

Horns and antlers are not the same. It is true that they both may be used for fighting, protection and display to attract a mate. However, that is where the similarity ends. Horns are made of bone that is covered with a tough protein based material. They usually begin to appear as the young animal approaches sexual maturity and are commonly found on both males and females. Horns typically continue to grow throughout the animal's life span and are nourished by major blood vessels. Antlers are usually found only in the males of species that produce them. They lack the same protective protein coat as horns and do not have blood vessels within them. Antlers are usually shed each year. The size of an antler indicates an organism's health, nutritional state and genetic dispositions, but not their age.

A Maggot with a Taste for Blood

In Africa, there are many types of pests and parasites. Larvae of the tumba fly have a particularly unsettling life cycle. At least five species of the fly have been described within the genus *Auchmeromyia*. The adults lay eggs upon damp soil, or some other suitable substrate. The fertilized eggs hatch to release larvae, more commonly called maggots. The maggots feed on the blood of numerous types of mammals by entering their bodies at night. Ordinarily, they enter through a wound in the skin. In the absence of such a wound, the determined larvae may crawl into the body through the nose or some other opening. At least one species has a taste for human blood. It is commonly called the "Congo floor maggot." It may enter a home through contaminated soil or hitch a ride on damp bedding or clothing that has been outside. The larvae often take on a red color after feeding within their hosts. Thereafter, they leave the host and drop out of sight to complete their life cycle. The maggots pupate and hatch into adult flies.

In the Nursery

The period of time between fertilization and birth in animals is often referred to as gestation. Gestation times vary widely among animals. Sometimes the term incubation is used for birds and other animals that keep their eggs in nests until hatching. Keeping in mind that a typical calendar year is 365 days in length, consider the average gestation times of some animals listed below. Very large mammals tend to have the longest gestation times. Some elephants have gestation periods that last 624 days on average. That of the sperm whale is 490 days; giraffes, 450 days. Camels typically have about a 370 day span. Blue whale gestation lasts about one year; the horse slightly less at 336 days. Average gestation time for the polar bear is 240 days; 257 for gorillas and 280 for the domestic cow. White tail deer have an average period of 210 days. Newborn sheep appear after about 154 days. Goats trail closely behind at 151 days and pigs follow with 115 days. Smaller placental mammals generally have shorter periods of gestation. Both house cats and dogs average about 63 days. Pet guinea pigs average a 68 day span of gestation, skunks 49 days. Rabbits reproduce very rapidly and efficiently. Their shorter 31 day gestation period helps to explain why. Mice and hamsters enjoy gestation terms of less than three weeks each. Mice average 19 days and hamsters 16. Marsupial mammals give birth to very tiny offspring that are often no bigger than a small insect. The young crawl to a pouch in the mother's body to feed and develop. Kangaroo babies emerge between 35 and 42 days after conception. Those of the North American Opossum are born in only 12 days. Incubation times of bird's eggs vary widely as well. The ostrich averages 42 days; the goose, turkey and peacock about 28 to 30. Chickens have an average nesting time of 22 days and pigeons may incubate their eggs from 10-18 days, depending on many environmental factors.

It is all in the Eye (brows)

The next time you see and recognize a person, pay close attention to the eyebrows of that person and to your own. Humans almost universally raise their eyebrows as a sign of recognition when they encounter one another. Many people remain unaware of this subconscious "eyebrow shift" or "eyebrow flash" as it is known.

Top Speeds

The fastest mammal on record is the cheetah. It can run for short bursts at speeds that reach or exceed 70 miles per hour. Some other top speed records are shown for comparison. Elephants can reach 25 miles per hour (mph); giraffes, 32 mph; rabbits, 39. Lions may reach 50 mph speeds to outrun zebras, a favorite prey item, which can only run 40 mph.

Numerous smaller zebra mussels growing on a native mussel shell. Courtesy of USFWS. Photographer not identified.

Unwelcome Visitor

Another example of the tremendous ecological problems that can arise from the introduction of foreign species involves the zebra mussel (*Dreissena polymorpha*). This mollusc is probably native to Russia and was accidentally introduced into the United States in the mid 1980s. The female mussels may produce close to one million eggs per year under ideal conditions. The tiny larvae that result from these fertilized eggs float upon the water and are thus able to disperse into new habitats. Young mussels become permanently attached to a substrate once they complete the larval stage. They cause tremendous problems in the United States. Countless amounts of money are spent due to their habits of clogging water pipes, drains, power stations and ship rudders. Heavy populations of these mussels also impart bad tastes and odors to the waters they invade. In many locations in the United States zebra mussels have competed with, and even completely wiped out, native molluscs.

Ivory

The term ivory has been used to describe many white colored, rock-like items throughout history. Today the term is most often reserved for the material making up the tusks of elephants and similar mammals. These tusks are actually made of modified dentine tissue, the layer just beneath the enamel of most organisms' teeth. Tusks are almost always derived from elongated, upper incisor teeth of mammals. Ivory was valued in times past for piano keys, jewelry and other decorative items. In addition to elephants ivory has been harvested from the tusks of the walrus, hippopotamus, narwhal and a few other organisms. The ivory trade made such a terrible impact on populations of various wild animals that it has been abolished in most parts of the world. A specialized type of ivory survives in a fossilized form from ancient mammoths and similar organisms. It is called odontolite and often has a blue-gray coloration due to changes in its chemical composition over time.

Social and Nonsocial Insects

Two orders of insects, Hymenoptera and Isoptera include species that biologists classify as eusocial. The term eusocial means "truly social." The insects in these groups display varying degrees of complexity in the social structure of their colonies. There is usually a single queen in each colony that is responsible for reproduction. All workers are female but are incapable of reproduction. All of their activities are geared toward feeding the queen and caring for her offspring. Male members of the colony are short-lived. Their primary mission is to leave their home colony and mate with a queen from another colony, thereby assuring a mixture of genes from one colony to another. Many species of termites, ants, honeybees, wasps, yellow jackets and other organisms have truly social colonies of this type. A number of species of hymenopterans do not spend their adult lives in organized colonies. They may be referred to as solitary for this reason. Examples include the potter wasp, tarantula hawk, mud dauber and cicada killer.

Down In the Dumps for So Long

The rate of breakdown for materials deposited into landfills, or otherwise discarded, varies widely. Degradation of these materials involves a number of complex chemical changes. The type of material is, of course, the primary factor. However size may also be important. Generally the more of a substance that is in contact with the environment, the faster it breaks down. So ten pounds of iron nails would break down more quickly than a solid ten pound block of iron. Food materials will almost always decay more quickly outside of plastic garbage bags than inside. Following is a list of known and estimated average degradation times for some common materials. Cotton and paper may take one to six months. Cigarette butts may take as little as one to five years to disappear from sight, but up to 25 years to completely degrade. A pair of discarded leather shoes could take as long as 25 to 40 years to break up. Metal food cans (often incorrectly called "tin cans") that are mostly made of steel, may take more than 50 years to disintegrate. Aluminum cans take around 80 to 100 years. Some scientists put the number as high as 300 years. Items made from plastic have not been widely available for very many years. It is difficult, therefore, to determine exactly how long they may remain in landfills. Some forms break down more quickly than others. It is clear that the breakdown of plastics is a very slow process. Small plastic film canisters remain around for 20 to 30 years. It is thought that other plastic products, such as reinforced squeeze bottles for ketchup or other materials, may take longer than 400 to 1000 years to disappear. Some scientists doubt that certain heavy-duty plastic containers will ever degrade. Compact disks, Styrofoam and polyester products may remain intact for more than 500 years. Best estimates for the breakdown of glass range from one thousand to more than one million years.

Some Hate It, Some Love It

The wild European boar, *Sus scrofa,* is a much discussed organism. They were introduced to the United States, perhaps as early as 1493 with Christopher Columbus. Populations are scattered throughout the Southeast. Outside of its native home, the wild boar is especially common in the Smoky Mountains. These organisms are believed to be descendants of two introductions of European wild boar that occurred within the past several decades. The wild boar may reach lengths of up to six feet. They are known for their ferocity, particularly when defending young. Most biologists tend to frown upon the presence of the wild boar in the Smokies. This non-native species often uses tusks on its lower jaw to dig for food. They destroy common and rare plants without discrimination. Scores of hunters are happy that this organism has become naturalized in the Smoky Mountains. It is prized by many for food.

Coal

Coal is a carbon-rich sedimentary material derived from plant sources. Under certain types of environmental conditions these plant materials slowly transform into a highly compacted material useful as a fuel. At one time coal was burned heavily as a fuel in the United States. It is still in use, though on a more limited scale. Today, very few homes utilize coal as a heat source. Peat is the poorest and least-processed grade of coal. It still contains a great deal of raw plant material. Lignite, also known as "brown coal," is the next highest grade. It retains a high percentage of water and releases volatile gases as it burns. Bituminous coal is the most widely used in the United States. It is a better heat source than lignite but produces much smoke. Finally, the most pure and highly prized form of coal is called anthracite. It releases a great deal of heat as it burns but produces only small amounts of soot and smoke.

Human Waste Factories

It has been estimated that the typical adult human generates more than one and one-half tons of trash per year, on average. This figure excludes garbage in the form of food that is usually broken down quickly by microbes and animals. Further, the average adult adds 50 gallons of sewage per day to the mix. Human sewage may harbor parasites and disease-causing organisms. Millions of dollars are spent in large cities to clean and treat sewage each day. Sewage treatment techniques vary according to locale. The terms "black water" and "gray water" may often be used in sewage treatment discussions. Gray water includes wastes from washers, kitchen and bathroom sinks, and the like. It can be carefully filtered to yield drinkable water. These filtration systems are most popular (and most often legally approved) in drought prone regions. Black water is this same material, plus human urine and feces. It is not usually thought of as being recyclable.

The Tail Tells the Tale

Biologists believe that wolves, wild dogs and domesticated dogs may convey their emotional state by way of their tails. A wagging tail is often regarded as a sign of contentment or happiness but may also signal uncertainty. If the tip of the tail is tucked deep between the legs, scientists think that the animal is attempting to show submission. Holding the tail parallel to the back may indicate a threat display. Raised hair on the tail, or elsewhere on the body, is usually interpreted as a sign of aggression or fear or as a threat display.

Bird Homing and Navigation

The ability of birds to find their way home after being displaced, and to navigate great distances during migrations, is still a bit of a mystery to biologists. No single mechanism appears to explain homing and navigation for all birds. Some species have been so well studied that biologists can manipulate their environment and predict their homing and navigation errors accurately. The abilities of other species have proven difficult to understand. Some birds undoubtedly utilize what is called "spatial sense" to help them find their home sites. They memorize familiar landmarks and their positions in relation to their homes. This fact may explain why many captive birds are unable to find their way back home if they are accidentally released into the wild. Other birds use the position of the sun to help them to orient their flight patterns. At night, some may utilize the positions of constellations as landmarks in the night sky. A few species of birds rely on odor patterns in their environments to help them orient toward their home sites.

Mimicry and Camouflage

Mimicry involves the imitation of one species by another. In Batesian mimicry, a harmless species mimics a harmful species. The harmful organism may release a toxin or cause some other sort of problem for a potential predator. Mullerian mimicry involves two harmful species that mimic one another. Other forms of mimicry have been catalogued as well. Examples of various types of mimicry follow. Camouflage coloration helps an organism to blend with its environment in order to evade predators or to help it capture prey. Chicks of the burrowing owl mimic the sound made by rattlesnakes when threatened. Certain species of caterpillars are pigmented and marked to blend with the twigs of woody plants. Some orchids have coloration patterns that superficially resemble particular species of insects. An organism called the "snowberry fly" has dark markings across its otherwise transparent wings. When viewed from behind, at a distance, the markings combine with other feature of the fly's body to mimic a spider. The sargassum fish is almost indistinguishable from the floating mat-forming seaweeds, known as sargassum, within which it lives. Many species of plants, particularly the orchids, display flowers that resemble female wasps and bees. They attract the attention of males that pollinate the plants. The oceanic butterfly fish has a spot at the base of its tail fin that resembles a large eye. The *Automeris* moth, a genus from Central America, has two such eye spots. They are displayed as the moth moves their front pair of wings forward. At least one species of caterpillar has an enlarged tail end that resembles the head of a snake, complete down to the eye spots and scale-like skin. An organism known as the "dead leaf butterfly" is able to blend, almost perfectly, with brown-colored leaves of certain plants.

Ants as Planters

Some species of ants and plants display an unusual symbiotic relationship that essentially involves gardening services in exchange for a nutritious meal. The wildflower *Sanguinaria canadensis*, commonly known as bloodroot, is a good example. These plants are typically found in enriched forest soils. Their creamy white flowers give rise to seeds that are partially surrounded by an oily mass known as an elaiosome. Some species of ants will harvest the seeds and carry them to their underground dens. The ants remove the elaiosome from the seeds and use it as food. The seeds, containing a new bloodroot embryo, are discarded by the ants in their waste sites. These waste sites are ideal sprouting and growing grounds for the plant.

Endangered, Threatened, etc. What Do They Mean?

Ecologists and the United States Government have developed a classification system pertaining to species of organisms that are in peril. In the first place, an extant species is one that is currently in existence. An extinct species is one that has completely disappeared from all its known habitats and is not surviving in captivity. On rare occasions, specimens of a species that has been declared extinct have been found alive in the wild. Sometimes a species is designated "extinct in the wild" if it is known only from captivity. Endangered species are those that scientists believe are headed toward certain extinction without intervention. The tag "critically endangered" expresses an even more impending peril. A threatened or vulnerable species is not classified as an endangered species but is in danger of becoming so due to things like continuing loss of habitat, disease, hunting and pressure from other factors in its environment.

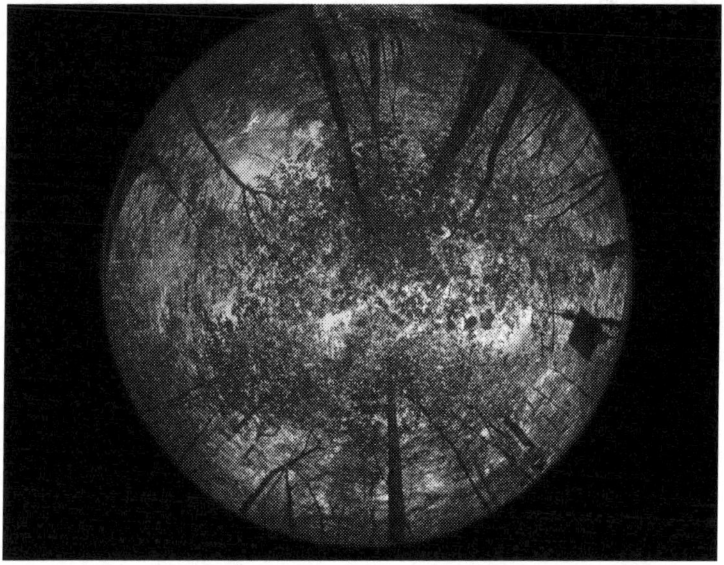

Looking up at the tree canopy, Eastern Deciduous Forest Biome. Courtesy of USGS. Photograph by Cynthia L. Cunningham.

What is a Biome?

Various communities of living organisms tend to exist within well defined geographic areas on the planet that are known as biomes. Two general categories of biomes exist. Those on land are called terrestrial biomes. Those within the water are known as aquatic biomes. Ecologists recognize a number of these biomes. Tundras, coniferous forests (also called boreal forests or taigas), grasslands, deserts, savannas, tropical rain forests, cloud forests, polar ice caps and temperate deciduous forests are examples of terrestrial biomes. The coniferous forests have evergreen cone-bearing trees such as pines, redwoods and hemlocks as the dominant tree types. Although these sorts of trees exist in temperate deciduous forests, these biomes are dominated by trees that seasonally shed their leaves. The term "deciduous" refers, in this case, to the act of shedding leaves. Tropical rainforests exist near the equator and regularly receive large amounts of rain each year. They have no winters and the temperatures stay within a fairly constant range. At especially high elevations some tropical rainforests are almost constantly wet and are called "cloud forests" by some biologists. Deserts are common along the equator as well but do spread further north and south than the tropical rainforests. They regularly receive little rainfall. In fact, some deserts may get little to no precipitation for decades. In terms of geographic locations and weather extremes, savannas tend to be intermediate compared to deserts and tropical rain forests. Grasslands occur in regions both north and south of the equator. As the name implies, the dominant plant material is grass. Trees tend to be few and extreme seasons exist. In the United States, the term "prairie" is often used to describe grasslands. Tundras are known to exist only toward the extreme northern areas of the globe. They have very cold temperatures and receive small amounts of rainfall. The soil is permanently frozen below a depth of about three feet and is therefore called permafrost. Above the permafrost the soil is marshy during warmer periods of weather. Plant life includes mostly grasses and a few shrubs. Aquatic biomes fall into two broad categories. Those with a high salt content are called marine biomes. They include saltwater lakes and seas, and intertidal zones. Many ecologists also recognize coral reefs, the oceanic continental shelves, continental slopes and the deep oceans as distinct biomes. Examples of freshwater biomes include swamps, streams and freshwater lakes. Estuaries are unusual aquatic biomes where a freshwater river empties into the sea.

What Should We Name the Baby?

Biologists and laypersons have come up with an interesting variety of names for juvenile animals. A small sample follows. Baby deer are known as fawns. Immature zebras, horses, mules and donkeys are often called foals. More specifically, foals are young males and fillies are young females. Spiderlings are the offspring of spiders. Most young fish are called fry or fingerlings; immature eels are known as elvers. A baby kangaroo, wombat, koala, opossum or most any other pouched mammal is known as a joey. Juveniles of the lion, cheetah, walrus, tiger, raccoon, hyena, leopard, and bear are called cubs. Immature frogs are called tadpoles. The term calf is used to describe the offspring of the rhinoceros, dolphin, whale, camel, cow, moose, hippopotamus, giraffe, elephant, bison and antelope. Puppy or pup is a term for dogs, bats, rats, mice, armadillos, hamsters, gerbils, moles, otters, wolves, beavers, sharks, squirrels, seals and sea lions.

More Babies Need Naming

A variety of terms may describe young birds including chick, fledgling and hatchling. More specific terms such as eyas for hawks, poult for turkey, squab for pigeon, gosling for goose, owlet for owl, duckling for duck, eaglet for eagle, cygnet for swan and diddle for chicken, are also used. Immature female chickens may, even more specifically, be called pullets and young male chickens cocks. Young sheep are called lambs or lambkins, while young goats are called kids. A piglet or shoat is a baby pig. A kit is a baby weasel or baby skunk. Bunnies are young rabbits. The term hatchling is used to describe young reptiles such as snakes, crocodiles, alligators and turtles. Snakes may be called snakelets. Young housecats, and the young of some other species of cats, are called kittens. The offspring of a platypus or echidna is sometimes known as a puggle.

Lousy and Fishy

Lice are not restricted to mammals and birds. Would you believe that fish can get lice? It is true. A genus of lice, *Argulus*, parasitizes a wide variety of fresh water fishes. The lice are large enough to be seen without magnification and are mostly green in color. They are flattened and attach to their hosts by way of hooks and suckers. *Argulus* species crawl all over the host fish's skin and may even enter their gills. The lice pierce the skin and gill tissue in order to feed on their host's blood. Moreover, they release a chemical to prevent the fish's blood from clotting during their feeding frenzy. Unlike many other kinds of parasites, these lice can swim easily from one host to another.

A Delicate Game of Give and Take

Several species within the genus *Tegeticula* play an intricate ecological game with plants in the genus *Yucca*. *Tegeticula* are yucca moths. They are the only organisms that are known to pollinate most species of yuccas, flowering plants that are widespread in the United States and elsewhere. Some species of yuccas would, in fact, be driven into extinction without the help of their pollinating moth friends. However the moths do not do their work out of a sense of altruism. They reap benefits from the relationship. A pair of yucca moths will mate within or near the yucca's large, showy flowers. The female will lay her eggs inside the blossoms. The mating and egg laying are accompanied by a meticulous transfer of pollen grains among the plants. Like other moths, the offspring of the insects will emerge as crawling larval caterpillars. During their development they will feed exclusively on the seeds of the yucca plant. Both the plant and the moth benefit from the relationship. The yucca plant does pay a price by giving up some seeds to the hungry larvae. However, no seeds would have been possible without the work of the moths.

What is Hibernation?

Hibernation is an extended period of time during which an organism undergoes very little to no activity. It typically takes place during the winter months in a number of small animals and is sometimes influenced by the severity of winter and duration of extremely cold temperatures. Prior to hibernation, the organisms involved typically store a great deal of body fat. Hibernation also involves such an extreme drop in body temperature that an organism may take hours or even days to return to normal activity. Many insects, snakes and other organisms hibernate. Small mammals such as bats, insectivores and rodents may hibernate routinely; others sometimes hibernate depending on environmental conditions. Contrary to what is often said large animals, such as bears, certainly do not carry out true hibernation. They may, however, undergo a state known as "winter sleep." These animals also prepare for winter sleep by storing fat. Their body temperature drops slightly. During winter sleep, animals are more easily aroused than a hibernating animal would be. In fact they may wake fairly often to feed, drink water and excrete wastes. In prolonged cold these activities may be suspended in some animals. Larger mammals that live in temperate regions often undergo winter sleep. Examples may include the black bear, grizzly bear, raccoon, beaver and many others. The term aestivation is often defined as "summer hibernation." Organisms that have a body temperature largely dependent on that of the external environment may burrow into the mud to escape the heat or to survive during dry weather. Some fish, frogs and other organisms aestivate during these types of circumstances. Some researchers claim a few mammals do so as well.

REFERENCES AND SELECTED READINGS

Abesha, E., Caetano-Anolles, G. & Hoiland, K. (2003). Population genetics and spatial structure of the fairy ring fungus *Marasmius oreades* in a Norwegian sand dune ecosystem. *Mycologia, 95,* p. 1021-31.

Abramoff, P. (1995). *Kingdom Monera.* W. H. Freeman Company.

Attenborough, D. (2002). *The life of Mammals.* Princeton University Press.

Baker, C. S. & Palumbi, S. R. (1994). Which whales are hunted? A molecular genetic approach to monitoring whaling. *Science, 265,* p. 1538-9.

Barbour, R. W. & Davis, W. H. (1969). *Bats of America.* University Press of Kentucky.

Barlow, C. (2002). *Ghosts of evolution: Nonsensical fruit, missing partners and other ecological anachronisms.* Basic Books.

Bart, J. (1995). Amount of suitable habitat and viability of Northern spotted owls. *Conservation Biology, 9(4),* p. 943-6.

Beacham, W., ed. (1997). *The official world wildlife fund guide to extinct species of modern times.* Beacham Publishing Group.

Bent, A. C. (1987). *Life histories of North American water fowl.* Dover Publications.

Benzing, D. H. (2008). *Vascular Epiphytes.* Cambridge University Press.

Bold, H. C. (1978). *Introduction to the algae: Structure and Reproduction.* Prentice Hall.

Brockman, C. F. (1986). *Trees of North America: A field guide to the major native and introduced species north of Mexico.* Golden Press.

Choe, J. C. & Crespi, B. J. (1997). *The evolution of social behavior in insects and arachnids.* Cambridge University Press.

Dahl, H. M. (1993). Things Mendel never dreamed of. *Medical Journal of Australia, 158*, p. 247-54.

Douglas, A. E. (2008). Mycetocyte symbiosis in insects. *Biological Reviews, 64*, p. 409-434.

Eisner, T. (2003). *For love of insects.* Harvard University Press.

Eldredge, N. (2002). *The patterns of evolution.* W. H. Freeman Company.

Evans, J. D., Shearman, D. C. A., & Oldroyd, B. P. (2004). Molecular basis of sex determination in haplodiploids. *Trends in Ecology and Evolution, 19,* p. 1-3.

Field, C. B., Behrenfield, M. J., Randerson, J. T. & Fallowski, P. (1998). Primary production of the biosphere: Integrating terrestrial and oceanic components. *Science, 281*, p. 237-40.

Fitzpatrick, J. W., Lammertink, M., Luneau, M. J. jr., Gallagher, T. W., Harrison, B. R. Sparling, G. M., Rosenberg, K. V.,

Rohrbaugh, R. W., Swarthout, E. C. H., Wrege, P. H., Swarthout, S. B., Dantzker, M. S., Charif, R. A. Barksdale, T. R., Remsen, J. V. jr., Simon, S. D. & Zollner, D. (2005). Ivory-billed woodpecker (*Campephilus principalis*) persists in continental North America. Science, 308, p. 1460-62.

Foelix, R. F. (2011). *Biology of spiders.* Oxford University Press.

Foote, L. E. & Jones, S. B. jr. (1998). *Native shrubs and woody vines of the Southeast. Landscaping uses and identification.* Timber Press.

Forbis, T. A. (2003). Seedling demography in an alpine ecosystem. *American Journal of Botany, 90,* p. 1197-206.

Galston, A. (1994). *Life Processes of Plants.* W. H. Freeman and Company.

Gende, S. M., Edwards, R. T., Willson, M. F. & Wipfli, M. S. (2002). Pacific salmon in aquatic and terrestrial ecosystems. *BioScience, 54*, p. 917-28.

Graham, L. E. (1999). *Algae.* Pearson Education.

Hall, E. R. (1981). *The mammals of North America.* Wiley Interscience.

Hanlon, R. T. & Messenger, J. B. (1996). *Cephalopod Behavior.* Cambridge University Press.

Hernandez, J. R. & Hennen, J. F. (2003). Rust fungi causing galls, witches' brooms and other abnormal plant growths in northwestern Argentina. *Mycologia, 95,* p. 728-55.

Hodgkison, R., Balding, S. T., Akbar, Z. & Kunz, T. H. (2003). Roosting ecology and social organization of the spotted-winged bat, *Balionycteris maculate* in a Malaysian lowland dipterocarp forest. *Journal of Tropical Ecology, 19 (6),* p. 667-76.

Holldobler, B. (1995). *Journey to the ants: A story of scientific exploration.* Harvard University Press.

Holloway, J. E. (2003). *Dictionary of birds of the United States.* Timber Press.

Hooper, J. N. & Soest, R. W. M. van., eds. (2002). *Systema porifera: A guide to the classification of sponges.* Kluwer Academic Publishers.

Jeanson, R., Duneubourg, J-L. & Theraulaz. (2004). Discrete dragline attachment induces aggregation in spiderlings of a solitary species. *Animal Behaviour, 67,* p. 531-7.

Jessop, B. M. (2003). Annual variability in the effects of, water temperature, discharge, and tidal stage on the migration of Americal eel elvers from estuary to river. *American Fisheries Society Symposium, 33,* p. 3-36.

Karol, K. G., McCourt, R. M., Cimino, M. T. & Delwiche, D. F. (2001). The closest living relatives of land plants. *Science, 249.* p. 2351-3.

Kartesz, J. T. (1994). *A synonymized checklist of the vascular flora of the United States, Canada, and Greenland.* Timber Press.

Kaston, B. J. (1978). *How to know the spiders, 3rd ed.* W. C. Brown.

Lee, D. S., Funderburg jr., J. B. & Clark, M. K. (1982). *A distributional survey of North Carolina mammals.*

Linsenmaier, W. (1972). *Insects of the world.* McGraw-Hill.

Linzey, A. & Linzey, D. W. (1972). *Mammals of the Great Smoky Mountains National Park.* University of Tennessee Press.

Margulis, L. & Schwartz, K. V. (1988). *Five kingdoms: An illustrated guide to the phyla of life on Earth, 2nd ed.* W. H. Freeman & Co.

Maton, A. (1997). *Parade of life: Monerans, Protists, Fungi and Plants, 3rd. ed.* Prentice Hall School Division.

McCafferty, W. P. (1981). *Aquatic entomology: The fishermans and ecologists guide to insects and their relatives.* Science Books International.

McCrea, K. D., & Levy, M. (1983). Photographic visualization of floral colors as perceived by honeybee pollinators. *American Journal of Botany, 70,* p. 369-75.

McLaachlan, A., Ladle, R. & Crompton, B. (2003). Predator-prey interaction on the wing: Aerobatics and body size among dance flies and midges. *Animal Behaviour, 66,* p. 911-5.

Moran, R. C. (2004). *A natural history of ferns.* Timber Press.

Moritz, R. F. A. & Neumann, P. (2004). Differences in nestmate recognition for drones and workers in the honeybee, *Apis mellifera. Animal Behaviour, 67,* p. 681-8.

Newsham, K. K., Rolf, J., Pearce, D. A., & Starchan, R. J. (2004). Differing preferences of Antarctic soil nematodes for microbial prey. *European Journal of Soil Biology, 40,* p. 1-8.

Nielsen, C. (2012). *Animal Evolution: Interrelationships of the living phyla, 3rd ed.* Oxford University Press.

Nsereko, V. L., Beauchemin, D. P., Morgavi, L. M., Roade, A. F., Furtado, T. A., McAllister, A. D., Iwaasa, W. Z., Yang & Wang Y. (2002). Effect of a fibrolytic enzyme preparation from *Trichoderma longibrachiatum* on the rumen microbial population of dairy cows. *Canadian Journal of Microbiology, 48,* p. 14-20.

Patterson, D. J. (1992). *Freeliving freshwater protozoa: A color guide.* CRC Press, Inc.

Pennak, R. W. (1978). *Freshwater invertebrates of the United States.* John Wiley & Sons, Inc.

Peterson, R. T. (1980). *Eastern birds, 4th ed.* Houghton Mifflin.

Prescott, G. W. (1978). *How to know the freshwater algae.* W. C. Brown.

Pollan, M. (2001). *Botany of desire: A plant's-eye view of the world.* Random House Publishing Group.

Popper, Z. A., Gurvan, M., Herve, C., Domozych, D. S., Willats, W. G. T., Tuohy, M. G., Kloareg, B. & Stengel, D. B. (2011). Evolution and diversity of plant cell walls: From algae to flowering plants. *Annual Review of Plant Biology, 62,* p. 567-90.

Primm, S. L., Jones, H. L., & Diamond, J. (1988). On the risk of extinction. *American Naturalist, 132.* p. 757-85.

Ramachandra, T. V., Mahapatra, D. M., Karthick, B. & Gordon, G. (2009). Milking Diatoms for Sustainable Energy: Biochemical Engineering versus Gasoline-Secreting Diatom Solar Panels. *Industrial and Engineering Chemistry Research, 48,* p. 8769-88.

Sherwood, G. D., Kovesces, J., Hontela, A., & Rasmussen, J. B. (2002). Simplified food webs lead to energetic bottlenecks in polluted lakes. *Canadian Journal of Fisheries and Aquatic Sciences, 59,* p. 1-5.

Shin, H. & Youn, J. (2005). Conversion of food waste into hydrogen by thermophilic acidogenesis. *Biodegradation, 16(1).* p. 33-44.

Smith, A. H. (1979). *How to know the gilled mushrooms, 2nd ed.* W. C. Brown.

Soltis, D. E., Soltis, P. S., Bennett, M. D. & Leitch, I. J. (2003). Evolution of genome size in the angiosperms. *American Journal of Botany, 90,* p. 1596-603.

Tiffney, B. H. (2004). Vertebrate disposal of seed plants throughout time. *Annual Review of Ecology, Evolution and Systematics, 35.* p. 1-29.

Volchan, E., Vargas, C. D., DaFranca, J. G., Pereira, A., & DaRocha-Miranda, C. E. (2004). Tooled for the task: Vision in the opossum. *BioScience, 54,* p. 189-94.

Wallraftt, H. G. (2004). Avian olfactory navigation: Its empirical foundation and conceptual state. *Animal Behaviour, 67*, p. 189-204.

Walter, D. & Procter, H. (1999). *Mites: Ecology, evolution and behavior.* UNSW Press.

Watling, R. (2003). *Fungi.* Smithsonian Institution Press.

Weber, N. S. & Smith, A. H. (1985). *A field guide to southern mushrooms.* The University of Michigan.

Wehr, J. D. (2002). *Freshwater algae of North America: Ecology and classification.* Elsevier Science and Technology Books.

Wofford, B. E. (1989). *Guide to the vascular flora of the Blue Ridge.* The University of Georgia Press.

Wofford, B. E. & Chester, E. W. (2003). *Guide to the trees, shrubs and woody vines of Tennessee.* University of Tennessee.